U0166190

Li Xinggang

李兴钢

著

Wandering, Walking, Viewing, Living

行者图语

浙江摄影出版社

脚、眼、脑、手

张永和

　　这个题目反映了建筑师李兴钢的这本书在我脑海里形成的第一印象，一个影像系列：

　　他的背影，他在行走中，双脚迅速移动的特写；

　　他的正面，他四处观望，眯缝着眼睛的特写；

　　他的侧面，他想到什么，头向下微低的特写；

　　他的膝部，上面摊开的速写本，手握着笔在纸上写画的特写。

　　过程中，镜头越推越近……

　　于是就出现了脚、眼、脑、手的系列：脚带动眼，眼刺激脑，脑指导手。

　　然而再一想，这个顺序并不准确。

　　人必须离开自己熟悉的环境去探查。但要走，要先看清方向。眼就在先了。

　　画其实是观察最好的方式之一，不画不会去注意许多事物。画帮助看，手帮助眼。手应在眼之前？

　　手脑是相互作用的，脑控制着手，手启发着脑。手脑不该分先后。

　　脑作为思考的器官，似乎统领着所有的身体机能，脑赋予脚走的动力、眼看的贪婪，手画的欲望，更应摆在首位。

　　脚、眼、脑、手完全可以倒转，手、脑、眼、脚也成立。或者，将这四个器官代表的行为分先后已失去意义。

　　还是一开始的那四个图像，只是现在都重叠在了一起。

　　从这个模糊不清的图像里，清晰地呈现出一个动态的人形，

Feet, Eyes, Brain, and Hands

Yung Ho Chang

This title reflects the first impression formed in my mind by architect Li Xinggang's book, consisting of a series of images:

His back—a close-up of his feet moving quickly while walking;

His front—a close-up of his eyes narrowing and looking around;

His profile—a close-up of his head lowering slightly while thinking;

His knees—a close-up of his hand holding a pen, writing and drawing on a sketchbook spreading out on his lap.

During this process, the camera lens gets closer and closer capturing feet, eyes, brain, and hands in such sequence: the feet lead the eyes, the eyes stimulate the brain, and the brain guides the hands.

With a second thought, however, I found that this sequence is not accurate.

People must leave their familiar environment to explore something different. But before leaving, one must first see the direction. Therefore, the eyes come first.

Drawing is actually one of the best ways to observe. Without drawing, one would not pay attention to so many things. Drawing helps looking, while the hands help the eyes. Should the hands come before the eyes?

The hands and brain are interacting in ways in which the brain controls the hands and the hands inspire the brain. Either the hands or the brain should not be prioritised.

The brain, as an organ for thinking, seems to dominate all bodily functions. The brain, which empowers the feet to walk with motivation, the eyes to look with greed, and the hands to draw with

他同时忙于跋涉、思考、眺望、记录，非常协调，非常熟练，又非常努力。他无意停下来，他是一台永动机，一台为了达到建筑创造佳境的永动机。

desire, should be given the priority.

Therefore, the order of feet, eyes, brain, and hands can be reversed, as the sequence of hands, brain, eyes, and feet also makes sense. Perhaps it is meaningless to separate the behaviors represented by the four organs.

Now, the four images mentioned in the beginning appear to be overlapped.

Within this blurry image, a dynamic human figure clearly emerges—he is preoccupied with trekking, thinking, overlooking, and simultaneously recording, with excellent coordination, skills, and diligence. He does not intend to stop. He is a perpetual motion machine, aiming to reach an optimal state of architectural creativity.

跋涉驻足，笔意双动；目及往还，心亦吐纳。

I believe that the way of experiencing and observing,
drawing and writing is both a response of reality to the ideal
and a demonstration of thoughts through memory.

行 / 旅

Walking

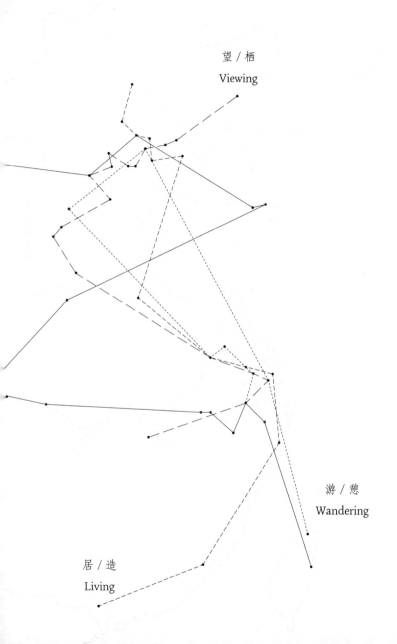

望 / 栖
Viewing

游 / 憩
Wandering

居 / 造
Living

前言

宋人丘葵有诗云："有美者杰阁，凌然跨虚空。有觉者彼岸，廓然大圆通。而我独得蠢，于何而适从。"许多年来，作为一个"没有地图的行者"，在抵达"廓然大通"之理想建筑的境地以前，在关于建筑的学习、思考和实践中，焦灼与低落的情绪时而有之。所幸天地广大，自然瑰奇，历史丰泽，生活辽阔，那些"阡陌纵横"的秘境，逐步将我引向另外一种自存的方式——正是于日常具体工作之外的"行、望、居、游"，这一大番"目既往还、心亦吐纳"所酿就的身体力行与身心感悟，将我从焦灼和低落中解脱出来，让我获得极为重要的思想源泉和启发，不断抽丝剥茧、拨云见日，直至形成今日愈渐清晰的"现实理想空间营造"的线路图。

"行、望、居、游"，代表了中国文化中的一种理想生活模式："行望"——在真实的天地山水中旅行，驻足眺望，格物致知，获得直观的人生感悟；"居游"——日常居住与卧游于山水自然之中，有如山居、村居，或将山水中获得的体验和场景浓缩和再造于自己的日常生活环境中，有如绘画、造园。郭熙《林泉高致》云："见青烟白道而思行，见平川落照而思望，见幽人山客而思居，见岩扃泉石而思游。"——在一个缩微的世界中，实现如此这般的愿望：通过日常生活中的"居游"而感知到由真实的天地山水中的"行望"才可获得的生命体验和内心感悟。这些"行望、居游"的历程和收获的感悟，被我以"草图"和微信式"短语"的方式记录下来，只因它们转瞬即逝，随时可能来，随时可能走，唯有描摹轮廓，演化字句，方可对这些令人怦然心动的"现场"予以现实的定格与捕获。我笃信，往来驻足，笔意双动，既

Introduction

The Song Dynasty poet Qiu Kui wrote that whereas a beautiful pavilion with maginificent spaces had a commanding presence under the sky and an intelligent man with transcendent awareness had a flexible attitude towards the world, what should I do given my slow-witted character? Over many years spent studying, thinking and practising in architecture, I have categorically abandoned the guidelines of ready-made maps and travelled without maps to explore an ideal architectural world in my mind. Throughout this long journey, endless happenings and thoughts, both complex and ambiguous, have caused me to become recurrently immersed in anxiety and weariness. Fortunately, the vast world, natural wonders, rich history and boundless life have all led me to another way of self-existence: the activities of wandering, walking, viewing and living outside my daily work. The feelings gained continuously through my observations of and engagement with nature have become an important source of thought and inspiration, freeing me from angst and despair. In the process of constant thinking and doing, step by step, the route map to "building realistic ideal spaces" has become clearer.

Wandering, walking, viewing and living represent aspects of an ideal lifestyle in Chinese culture. Walking in and viewing a natural landscape, investigating and understanding things, help me gain direct insights into life. "Living" and "wandering" here can refer to the activities of daily living in mountain homes or villages, or condensing the experiences obtained or the scenes framed in the natural landscapes and re-creating them in the daily environment which is a process similar to painting or gardening. In the book *Linquan gaozhi*, the painter Guo Xi wrote that "a good landscape painting is a place that people would aspire to walk after seeing the painted smoke and path, to view after seeing the plain and setting sun, to live after seeing hermits and to travel after seeing rock and spring." I intend to realise such a desire in a miniature world.

是彼岸之于此岸的召唤，亦是记忆之于思想的观照。

谓之"草图"，乃因我所绘并非为"美"或"构图"的美术画作或写生，而是我对所观想之空间、景物乃至画作、思想有所感悟启发时的快速即时记录，因此我更愿称之为"图"而非"画"，非经由观察体验而最得触动及启发者不动笔也；又因速写之"速"而草图之"草"，重点在于关键要害而非求全记录，并由此留下不可磨灭的印象，无须担忧其潦草不美，价值仅在领悟之记录、之提醒、之反思也——中国的山水画作亦常常被称为"图"，当与其思想性、记录性而非仅限于"美术性"甚或"写真性"相关。

相比文字而言，"图"对于空间营造的记录自然具备天然的功能优势，但短小精悍的文字之"语"并不可缺，是对"图"的补充和提炼，是思考的记录，图语对照，方成整体——犹如中国古代文人画者的行旅记游之册页，往往诗画对照，方成品题。

自2013年重拾学生时代的习惯，以"图语"记录行旅感悟，迄今已逾六年，此六年同时亦是关于"胜景几何"的思考和实践之关键时期，今日小结，竟得图400余幅。相信自己从中获益良多，虽然并不能道清其中所有具体关联，以及某些触动与启发是否或将作用于未来："地图与旅途"——思考与实践可以一直相互映照、批判、修正，不断纳入生机勃勃的现实，拂晓实践的出路，照亮思想的盲区，觉察思考的误点，调整行动的方向，朝着超越个人与文化的、应许于人类生存与生活的理想建筑世界，一直行走下去。

从所有草图中挑选并辑此小册，与我平日用的笔记本大小相仿。若按原来的"制图"时间顺序自然编排，对于读者而言，唯

The life experience and inner feelings obtained from walking and viewing in nature can be perceived in everyday living and travelling activities. The sketches and WeChat-style phrases presented in this book record these experiences of wandering, walking, viewing and living and the associated feelings. Because they are fleeting, and may come and go at any time, only by portraying their profiles and depicting them with words can we record and capture these fascinating "moments" realistically. I believe that the way of experiencing and observing, drawing and writing is both a response of reality to the ideal and a demonstration of thoughts through memory.

These so-called sketches are not artisans' paintings or sketches highlighting beauty or composition. They are a quick and immediate record of spaces, scenery, paintings or thoughts that have impressed me or inspired my thinking. Therefore, I would call them "drawings" rather than "paintings", and I would not draw them without first-hand observation and experience. These drawings emerged from inspiration and impulse. Because of the speed and hastiness of sketching, the key aim is not to seek a full record, but to capture the most crucial component that left an indelible impression. There is no need to worry about whether the sketches are beautiful, because their value is embodied in the understanding they remember, record and reflect. Chinese landscape paintings have often been called drawings, mainly because of their ideological, documentary nature. They are not limited to artistic or even realistic characteristics.

Compared with text, drawing has a naturally functional advantage in recording spatial construction, although short and precise phrases are also indispensable. Text is a supplement to and refinement of drawing and the record of thought. Together, drawings and phrases are a whole, similar to the travelogues of ancient Chinese literati painters, which often combined poetry and

恐过于零散随意，甚至还有反复之处（如颐和园，不同时间和季节去过多次）。因此，特按"游、行、望、居"，分别以"憩、旅、栖、造"相叠加，成游／憩、行／旅、望／栖、居／造，分类次第编排，对应园林与仿画、山水与旅行、聚落＋城市＋建筑、日常＋场地＋设计＋研究，既体现外在观照对象的不同，又暗含内在思想、行动的关联和发展。

是谓"行者图语"。

paintings.

It has been more than six years since 2013, when I regained the habit of my student days and began to record my impressions about travelling through sketches. The last six years have also been a crucial period for me to think about and practice the discourse of Integrated Geometry and Poetic Scenery (*Shengjing Jihe*). So far, I have produced more than 400 sketches. I believe that I have benefited a lot from them, and although I cannot discern all of the specific connections, and whether some effects and inspirations may appear in the future. Both maps and journeys, and thoughts and practices, can always be mirrored, criticised, and revised. It is my hope to constantly incorporate the vibrant and diverse reality, to discover the highlights of practice, to illuminate the blind spots of thought and to continuously walk towards the ideal architectural world of human existence and life that transcends individuals and culture.

I have selected my sketches and compiled them into this booklet, the size of which is similar to my daily notebook. However, I have not collected them in the chronological order of their original production. This would be too scattered, random and even repetitive for readers. Some of the sketches, such as those of the Summer Palace in Beijing, were drawn in one place but at different times and in different seasons. Therefore, the book has been divided into four sections: wandering, walking, viewing and living. These sections correspond respectively to the gardens and imitations of noted paintings; landscapes and travelling; settlements, cities and buildings; and daily life, sites, designs and research. This categorisation not only reflects the differences in external objects, but also implies the relationships between and development of internal thoughts and actions. These are so called a walker's illustrated essays (*xingzhetuyu*)—wandering, walking, viewing, living.

游 / 憩（园林与仿画）

　　中国园林是对"现实理想空间营造"最具启发性和标本性的"模型"。我对园林的兴趣、思考与实践由来已久，若从2000年北京兴涛展示中心的设计开始算起，已近20年。无论苏、杭、扬州一带的江南园林，还是广东、广西一带的岭南园林，北京、承德、山西一带的北方园林，都有涉足。对中国园林的种种体验和感悟，诸如造园的要素（围墙、山石、池塘、林木、建筑）、手法（大小、高下、虚实）、境界（如童寯先生所言：疏密得宜、曲折尽致、眼前有景），而其中的造园情趣与生活哲学，对我今天的建筑思考有着根源性的影响。本册收录近年所作部分园林图语，聊为代表。

　　中国园林由"山水别业"发展到"城市山林"，是对自然山水的缩微与仿造，所谓"移天缩地""模山范水"，并由园林及至盆景、玩石，其意象境界之营造受中国山水画作影响甚多。因此有机会面晤研读一些重要及喜欢的传统山水画原作，我亦尝试仿写它们，希望借此"实地"想象和领略其中的图绘手法、空间布局、景物层次、意象境界，并与园林的营造相互对应参照。

Wandering
(Gardens and imitations of noted paintings)

Chinese gardens are both the most inspiring and the standard model for "building realistic ideal space". I have been interested in thinking about and designing gardens for nearly 20 years, since the design of the Beijing Xingtao Exhibition Centre in 2000. I have visited and studied the Jiangnan gardens in Suzhou, Hangzhou and Yangzhou, the Lingnan gardens in Guangdong and Guangxi provinces and the northern gardens in Beijing, Chengde and Shanxi province. My various experiences and perceptions of Chinese gardens, such as the elements (walls, stones, ponds, trees, buildings), techniques (big versus small, high versus low, void versus volume), realms (as Tong Jun claimed, appropriate sparseness and density, twists and turns and dynamic scenery), particularly the temperament and interest (*qingqu*) and the philosophy of life represented in gardening, have taken root to influence the formation of my architectural thoughts today. This book is a collection of sketches and phrases depicting some of the gardens I have drawn and written about in recent years.

As microcosms and emulations of natural landscapes, Chinese gardens have evolved from "mountain and river villas" (*shanshui bieye*) to "urban scholarly gardens" (*chengshi shanlin* or Rus in urbe). The imagery and conceptional construction of the so-called "moving of the sky and shrinking of the earth" (*yitian suodi*) and "imitating mountains and rivers" (*moshan fanshui*) are both influenced by Chinese landscape paintings. I have had the opportunity to encounter and study some important and favourite traditional landscape paintings. I also try to imitate them, to imagine and appreciate the paintings' techniques, spatial layout, multi-layered scenery, images and conceptions, and use them as cross-references for garden-making.

1-001 仿写 [明] 唐寅《湖山一览图》 （2014.12.10）

深远之作。远，湖舟浮行；近，丛树、巨石、书房、木桥、水岸；中，悬壁、礁石、树中亭、崖顶房廊中静坐之人——湖山一览者。所有画面的空白图底成为观者想象中的洋洋湖面——非水之水，无材之材，谓之抽象。

1-001 Sketch imitating Ming Dynasty painter Tang Yin's *Overview of Lakes and Mountains* (*Hushan yilan tu*) (10 December 2014)

This is a work with a sense of deep distance: a boat floating on the lake in the far distance; bushes, boulders, a study room, a wooden bridge and the waterfront in the foreground; and in the middle, cliffs, reefs, a pavilion among the trees, and people sitting in a house on the cliff top gazing at the lake and mountains. For audiences, the blank bottom of the painting is imagined as the wide lake surface, the water without water, the feature without feature, creating a sense of abstraction.

〔明〕唐寅 湖山一览图（轴）2014.12.10.

中国美术馆藏

1-002 (1-4) 故宫，石渠宝笈特展，速写 [清] 王原祁《西岭春晴图卷》（2015.09.28）

长卷，自右至左徐徐展开。每一局部皆可独立成图，所有局部构成连绵统一的整体。远山近岸，高岭深谷，危崖礁石，丘陵岛渚，宽滩长堤，老树嫩丛，木桥石板，山寺村居，石亭水榭，春水晴日，宁致人生。

1-002(1-4) Special Exhibition of *Shiqu baoji* in the Forbidden City, Beijing. Sketch of the *Scrolls of Xiling in the Clear Spring* (*Xiling chunqing tujuan*), originally drawn by Wang Yuanqi of the Qing Dynasty (28 September 2015)

The painting is a long roll, slowly unfolding from right to left. Each part of the painting can be seen as independent, but together all of the parts form a continuous unified whole. The elements include distant mountains, near shore, high ridges, deep valleys, towering cliffs, reefs, hilly islands, wide beaches, long dikes, old trees, tender groves, a wooden bridge with stone slabs, temples and villages within the mountains, pavilions beside stone and water, water and sunny days in spring and a sense of tranquillity.

西鎮春晴
做大痴筆以
帕春蕪直
陌城夏日子來
偶眾絵中
崔宸

2015-09-28　〔清〕王原祁西鎮春晴圖卷
（之一）

1-003 故宫，石渠宝笈特展，速写 [宋] 夏圭《梧竹溪堂页》（2015.09.28）

溪边歇山草堂，梧竹幽居。"……岩扉暖护青萝幕，涧水凉分白芷衷。竹化白龙苍雾湿，池藏金鲤锦萍收……"

1-003 Special Exhibition of *Shiqu baoji* in the Forbidden City, Beijing. Sketch of the *Wuzhu xitang ye*, originally drawn by Xia Gui of the Song Dynasty (28 September 2015)

A pavilion with a gable and grass-covered hip roof is located by the riverside, and a secluded house is surrounded by bamboo. A poem written on the side of the painting describes the scenery of a pavilion with stone, Chinese usnea, a mountain stream, bamboo, smoke, a pool, fish and lotus plants.

29

2a5, 09. 28

記得當年故國遊風主

天際碧雲秋岩飛暖籠

青蘿幕洞水涼分白芋

裳竹化白鷺鷥竊務沒

池藏金鯉錦詳妝不知

近日吟壇支清興還如

社裏不

越東桂童

选自 [宋] 夏主 梧竹溪童久

1-004 故宫，石渠宝笈特展，速写 [宋] 马远《梅溪放艇页》
（2015.09.28）

远峰巍峨，梅花树下，山溪岸边，小艇浮水，惬意自然。

1-004 Special Exhibition of *Shiqu baoji* in the Forbidden City, Beijing. Sketch of the *Meixi fangting ye*, originally drawn by Ma Yuan of the Song Dynasty (28 September 2015)

A towering peak at a great distance, a small boat floating on the stream near a plum tree. What natural and cosy scenery!

马远
(宋) 梅溪放艇页

2015·09·28

1-005 台北故宫博物院，仿写 [明] 董其昌《夏木垂阴图》

（2016.01.24）

冬季再到台北，来看雨。夜访董其昌，妙在能合，神在能离。董的中部山石画法似是对前人画石的"再画"，一种抽象化的发展与加工。

后来（2018.12.27 下午）在上海博物馆观董其昌书画艺术大展"丹青宝筏"，当晚在上海那行零 == 回度空间讲座及讨论佛光寺，两者巧合，感想良多。前者是如何师法古人与造化，后者是如何对待历史与当代，类似的创作焦虑和努力始终甚至永远存在；如果把"古人"看作狭义回的"历史"，那么"当代"即是广义的"造化"；这种"造化"其实由多样的、立体的"历史"经历时光层层叠加而成（历史就是"经历"的史），因此理论上所有的"当代"都有自己的特别之处，只不过佛光寺是靠近极致的一个特别案例；另外对于实践者来讲，思想和"宝筏"同样重要；还有对于男性创作者来说，通常要六十岁以后才能绽放，所以不必着急（龇牙笑）。

1-005 The National Palace Museum in Taipei. Sketch imitating Ming Dynasty painter Dong Qichang's *Shady Trees in a Summer Landscape* (*Xiamu chuiyin tu*) (24 January 2016)

I went to Taipei in the rainy winter and visited an exhibition of Dong Qichang's work at night. It was characterised by "marvelous at synthesis, divine at departure". Dong's method of painting stone, demonstrated in the scroll's centre, seems to be a "reproduction" of his predecessors' work, a development and process with abstraction.

Later, in the afternoon of 27 December 2018, I visited Dong Qichang's exhibition, *The Ferryman of Ink World* (*Danqing baofa*) at the Shanghai Museum. That evening, I gave a lecture on the Foguang Temple and took part in a discussion. The two events coincided and made me feel a lot of emotion. The former was about how to learn from the ancients when creating; the latter was about how to treat history and contemporary life. Similar anxiety and effort in creating have always existed, and will always exist; if the ancients are regarded as history in the narrow sense, then contemporary life is about "creation" in the broad sense. Actually, such a "creation" is produced through the juxtaposition of multi-layered, three-dimensional historical experiences over time (history is the history of experience). Therefore, in theory, all of contemporary life has its own singularity. However, the Foguang Temple is an extraordinary special case. For practitioners, thought is as important as technique (*baofa*). In addition, male creators usually reach maturity at the age of 60, so it is not necessary to be anxious!

1-006 (1-10) 台北故宫博物院，仿写 [元] 黄公望《富春山居图》（《剩山图》《无用师卷》），"山水合璧"（2016.01.24—02.07）

看到《富春山居图》，方知《西岭春晴图卷》定是王原祁在向黄公望学习和致敬。远、中、近，高、深、平，如波浪般连绵不断地涌现。

1-006 (1-10) The National Palace Museum in Taipei. Sketch imitating Yuan Dynasty painter Huang Gongwang's *Dwelling in the Fuchun Mountains* (*Fuchun shanju tu*) (*shengshan tu, wuyongshi juan*, the two parts of the painting are perfectly combined) (24 January to 7 July, 2016)

When I saw *Dwelling in the Fuchun Mountains*, I realised that in *Xiling chunqing tujuan*, Wang Yuanqi had definitely learned from and paid tribute to Huang Gongwang. With far, middle and nearby views, and high, deep and level distances, with the scene dominated by mountains, it was similar to waves, continuously emerging.

2016.01.24 宏字墅平昌夏茉野陰圈台北

1-008 台北故宫博物院，仿写 [清] 王原祁《仿黄公望山水》

（2016.02.14）

仿而仿之，有趣的经验。

1-008 The National Palace Museum in Taipei. Sketch imitating Qing Dynasty painter Wang Yuanqi's *Imitating Huang Gongwang's Landscape Paintings (Fang huang gongwang shanshui)* (14 February 2016)

It is an interesting experience to duplicate an imitated painting.

摹 [清] 巴慰祖 秋山图   2016. 2. 14.

2016.02.14. 菁菁 川清 王厚祁 仿黄公望山水

1-009, 010 仿写山水 (2017.02.01)

均为深远构造。对页图左山层层，由近推远，近水流石，山台亭树；下图山作环围，空间分出前后，中为焦点小峰，人、亭其上。

1-009, 010 Sketches imitating the landscape (1 February 2017)

Both paintings include the technique of deep distance. The left part of the painting on page 45 features multi-layered mountains, with a river and stones in front and mountains, platforms, pavilions and trees in the distance. The painting below depicts the mountain surrounding the valley, dividing the foreground and background, where in the middle there is a small peak with people and a pavilion on the top.

1-011

游瞻园，速写 [清] 袁江《瞻园图》（2015.07.22）

　　"六朝如梦鸟啼花，况复中山魏国家。今日瞻园吊遗迹，只余残石数堆斜。"园林代代兴衰更替，如生命循环，故事与园图，遗迹与新造，交杂生动。大假山应始终为园中枢纽主体，古意可与今通感，只不过换了叠山匠师（敦桢先生）。城市山林，与城外山水寺塔应和一体。

1-011 Visiting Zhan Garden. Sketch of the Qing Dynasty painter
Yuan Jiang's *Zhan Garden* (22 July 2015)

Viewing the ruined Zhan Garden, experiencing historical change.
With its remaining stones, one may have a feeling of nostalgia.
The gardens, like life, have a circle of regeneration. The stories and
garden-drawings, relics and new creations are vividly mixed. The
big rockery should always be the hub and main body of the garden,
whose ancient meaning is connected with the present situation, but
with a different rockery designer and builder (Liu Dunzhen). The
urban scholarly gardens are integrated with the temple, tower and
landscape outside the city.

（摹）清·袁江《瞻园图》之右幅
2015-07-22

1-012 苏州返京高铁，仿写 [清] 刘懋功《寒碧山庄图》

（2019.05.25）

1-012 On the high-speed train from Suzhou to Beijing. Sketch
imitating Qing Dynasty painter Liu Maogong's Hanbi Mountain
Villa (*Hanbi shanzhuang tu*) (25 May 2019)

2019.05.25. 摹刘懋功 1857《寒碧山庄图》(之一) 田部

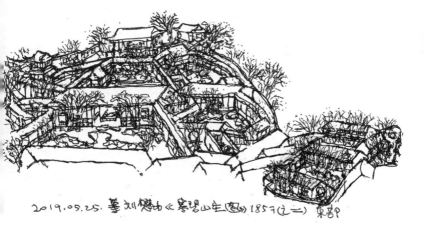

2019.05.25. 摹刘懋善《寒望山生（色）》1857(之一) 东都

1-013 摹 [清]《三山五园图》 （2019.06.30）

1-013 Sketch imitating The Three Hills and Five Gardens (*Sanshan wuyuan tu*) (30 June 2019)

1-014 杭州陆羽山庄，摹写径山禅寺 (2016.03.14)

"径山乃天下奇处也，由双径而上，至绝高之地，五峰巉然。中本龙湫，化为宝所……特为伟异，天作地藏，待斯人而后发。"

1-014 Hangzhou, Luyu Mountain Villa. Sketch of Jingshan Temple (14 March 2016)

Jingshan is an uncanny place. There are two paths leading to the top, where five peaks stand. The central peak should be a pool inhabited by a dragon (*longqiu*, or the residence of dragon), becoming a precious place. This great and singular place created by nature is waiting for people to interpret in the future.

径山禅寺

2016.03.14 摹写 于陆羽山庄 杭州

1-015 颐和知春画中游，冰风吹脸笔头僵 (2016.02.13)

知春亭望万寿山、玉泉山。

1-015 The Summer Palace, Zhichun Pavilion and Huazhongyou (a place called Strolling through Painted Scenery). The icy wind blows on my face and my pen becomes stiff (13 February 2016)

Standing in Zhichun Pavilion, I look at Wanshou Hill and Yuquan Hill.

2016.02.13. 颐和园知春亭

1-016 颐和知春画中游，冰风吹脸笔头僵（2016.02.13）
画中游平台望玉泉山。

1-016 The Summer Palace, Zhichun Pavilion and Huazhongyou. The icy wind blows on my face and my pen becomes stiff (13 February 2016)

Standing on the platform of Huazhongyou, I look at Yuquan Hill.

1-017 冰上颐和 （2017.01.27）

1-017 Frozen Summer Palace (27 January 2017)

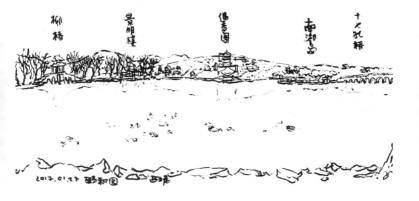

1-018 颐和园西堤，"树桥"（2017.01.27）

1-018 West bank of the Summer Palace, "tree bridge" (27 January 2017)

2017.01.27. 颐和园 西堤 树桥

1-019 冰上颐和 （2018.01.01）

1-019 Frozen Summer Palace (1 January 2018)

2018.01.01 颐和园

1-020 颐和园南如意门绣漪桥边 （2018.01.01）
深远，层出不穷。

1-020 South Summer Palace, Ruyi Gate, Xiuyi Bridge (1 January 2018)

Deep distance, with endless layers.

1-021 春雪颐和：十七孔桥与南湖岛 （2019.02.12）

初次试用毛笔水墨草图，感觉不同。

1-021 Snowy Summer Palace in spring: the 17-arch Bridge and Nanhu Island (12 February 2019)

The first time I tried to use a brush and ink to sketch, feeling the difference.

1-022 春雪颐和：十七孔桥（2019.02.12）

1-022 Snowy Summer Palace in spring: the 17-arch Bridge (12
February 2019)

2019.02.12. 颐和园十七孔桥，雪后

1-023 春雪颐和：雪中远望廊如亭、十七孔桥与南湖岛
（2019.02.12）

1-023 Snowy Summer Palace in spring: overlooking Kuoru Pavilion, the 17-arch Bridge and Nanhu Island (12 February 2019)

2019.02.12 颐和园 十七孔桥与南湖岛，廊如亭，雪中

1-024 春雪颐和：佛香阁东望转轮藏（2019.02.12）

1-024 Snowy Summer Palace in spring: looking east to Repository of Sutras (*Zhuanlunzang*) from Foxiang Pavilion (12 February 2019)

2019.02.12 颐和园佛香阁东望转轮藏 雪后

1-025 春雪颐和：佛香阁前大台阶西望玉泉山和玉峰塔

（2019.02.12）

1-025 Snowy Summer Palace in spring: looking west to Yuquan Hill and Yufeng Tower from the front step of Foxiang pavilion (12 February 2019)

雪后
2019.02.12 颐和园佛香阁大台阶 西望 玉泉山 玉峰塔

1-026 春雪颐和: 佛香阁西望宝云阁和玉峰塔 (2019.02.12)

1-026 Snowy Summer Palace in spring: looking west to Baoyun Pavilion and Yufeng Tower from Foxiang Pavilion (12 February 2019)

2019.02.12 佛香阁 西望 宝云阁. 2.峰塔. 雪后

1-027 春雪颐和：佛香阁远眺南湖岛，平远山水（雪）
（2019.02.12）

1-027 Snowy Summer Palace in spring: looking north to Nanhu Island from Foxiang Pavilion, level distance of landscape (snow) (12 February 2019)

2019.02.12. 颐和园 佛香阁远眺 南湖岛. 还远山水画意

1-028 春雪颐和：树石之上的智慧海（2019.02.12）

1-028 Snowy Summer Palace in spring: a Buddhist Temple (*Zhihuihai*) built above stone (12 February 2019)

2019.02.12 雪后颐和园，翔石之上的智慧海

1-029 春水颐和：像山一样的水（波）——"山水一体"

（2019.03.17）

1-029 Summer Palace water in spring: water (waves) like mountains
— Integrated "mountain" and water (17 March 2019)

2019·03·17·

1-030 春水颐和：细风吹皱水面 （2019.03.17）

1-030 Summer Palace water in spring: mild wind wrinkling the water's surface (17 March 2019)

2019.3.17 昆水颐和（细风吹皱水面）

1-031 颐和园，丁垚老师现场课。后山落水，须弥灵境（右）西侧（2019.03.29）

1-031 Summer Palace. Lectured by Ding Yao on the site of the falling water through rocks at the back hill, west side of Xumilingjing (right) (29 March 2019)

颐和园后山落水
(须弥灵境旁侧)
(西)

2019.03.29.

1-032 "山色因心远，泉声入目凉"：宝云阁庭院，山体上的曼陀罗理想世界（2019.03.29）

1-032 The mountain seems very far from the heart, and I feel cool after hearing the sound of mountain springs flowing: the Baoyun Pavilion courtyard, the ideal world of the mandala on the mountain (29 March 2019)

云南朝院：叠山置阁随地势山依上的

1-033 颐和园后山，赅春园遗址（2016.05.01）

"萝径因幽偏得趣"：乾隆这个人，实在是真爱玩、"城会玩"啊。

1-033 The back hill of the Summer Palace, the Ruins of Gaichun Garden (1 May 2016)

The tranquil path with vines is interesting (*luojing yin you pian dequ*): Emperor Qianlong really loved to play, and indeed was able to play.

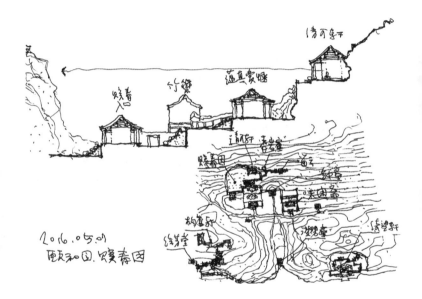

2016.05.01
颐和园.谐趣园

1-034 赅春园遗址，西桃花沟上钟亭及跨沟平台遗迹

（2016.05.02）

沟侧林中，东望赅春园，最简庭院，却格外幽深，"山阴最佳处"。弘历《赅春园》诗句："山阴或不来，来必憩斯轩。"

1-034 The Ruins of Gaichun Garden, the Zhong Pavilion nearby on West Peach Blossom Ditch and the cross-ditch platform ruins (2 May 2016)

Standing in the forest at the side of the ditch, I look east at Gaichun Garden, the simplest courtyard, but especially tranquil, the best place under the shadow of the hill. The verse about the Gaichun Garden written by Hongli (Emperor Qianlong) shows that the shadow of a hill may not appear; if it comes, it must be cast on the pavilion.

2016.5.2. 畅春园遗址

1-035 赅春园遗址，香岩室西侧留云内后墙岩壁摩崖遗迹

（2019.03.17）

乾隆《清可轩》诗云："恰当建三楹，石壁在其腹。"墙下有房，房中有石，石中居佛，栖岩幽趣。

1-035 The Ruins of Gaichun Garden, the cliff ruins in the back wall of Liuyunnei on the west side of the Xiangyan Room (17 March 2016)

Emperor Qianlong's poem, *Qing Ke Xuan* writes that inside of the three-room building is a huge stone like a cliff or wall. What is interesting is the way in which a Buddha was cast into and is inhabiting a rock within a house adjacent to the wall.

2019.3.17. 嫩春园遗址

塔下有房，房中有石，石中居佛

1-036 香山见心斋轴测示意图 （2018.02.22）

　　明嘉靖元年（1522 年）建，清嘉庆元年（1796 年）重修。小小斋园，跨越明清两代近 300 年，中有见心斋、知鱼亭、正凝堂、畅风楼等，南北二门居于两侧，沿山坡的曲线围墙，以为特色。

1-036 Axonometric diagram of Jianxin Room at Fragrant Hill (*Xiangshan*) (22 February 2018)

　　It was built in 1522 during the Jiajing period of the Ming Dynasty, and was rebuilt in 1796 during the Jiaqing period of the Qing Dynasty. The small Jianxin Room, spanning the Ming and Qing dynasties for nearly 300 years, consists of a group of buildings such as Jianxin Room, Zhiyu Pavilion, Zhengning Hall and Changfeng Tower. With the north and south doors are on both sides, Jianxin Room features curved walls along the slopes.

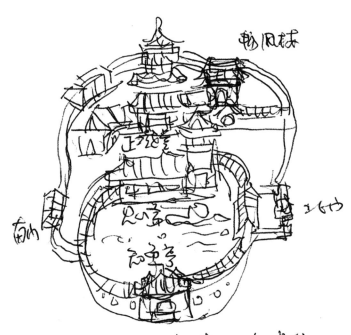

畅风楼

正觉寺

颐和

2.5.4

青山忆山斋
2018.12.23
道

明嘉靖1552年建成
清嘉庆1796重修

87

1-037 殿宇深处有花园 （2018.03.15）

　　故宫建福宫花园，据称后来在宁寿宫花园（乾隆花园）中被称为"迷楼"的符望阁的建筑形制即是仿自此园中的延春阁。

1-037 A garden in the depth of the palace (15 March 2018)

　　It is said that the building form of the Fuwang Pavilion, which is known as *mi lou* in the Ningshou Palace Garden (Qianlong Garden), originated from the building of the Yanchun Pavilion in the Jianfu Palace Garden in the Forbidden City.

2018.03.15 古色建筑故宫花园

1-038 晋祠难老泉真趣亭 (2016.06.09)

"穿花蛱蝶深深见。"（杜甫）难老泉真趣亭位于献殿、鱼沼飞梁和圣母殿三国宝构成的中轴之侧，晋水源初出石塘拐角水面放大处，亭在北岸，临流制高，是一处下沉式园中之园。其入口竟在真趣亭下拱形石洞处，由洞内石梯进入，豁然可见水流潺潺，曲桥横浮；西岸难老泉水注入，众人嬉泉水抚僧像头，岸上游人俯瞰；水边上岸，缘渠而南行，则见山林幽深。亭的基址，原仅仅是个石梯口，是村人为洗纸方便而开凿，后来一位在此驻军的旅长发现，晋祠宇下，饶有真趣，便建此亭。不论今古，勿论文武，有真趣可也。若干年前阿瑜曾带我游览晋祠，却未注意此佳妙处，推之荐之。

1-038 The Zhenqu Pavilion of the Nanlao Spring, Jinci Temple (9 June 2016)

Butterflies flitted through the flowers, appearing and disappearing (chuanhua jiadie shenshen jian) (written by Tang Dynasty poet, Du Fu). The Zhenqu Pavilion of the Nanlao Spring is located on the side of the central axis of three treasures: Xian Temple, Yuzhao Feiliang and the Shengmu Temple. The source of the Jin River flows from the corner of the pond with the rocky shore. The waterside Zhenqu Pavilion, within a sunken garden on the north bank but on a high place, is accessible through an arched stone cave. When entering by the stone ladder inside the cave, one suddenly sees the flowing spring and the floating curved bridge. The Nanlao Spring flows from the west bank into the pond while tourists on the shore overlook that people are paddling and touching the head of the statue. Walking along the edge of the ditch, one can see deep and secluded woods. The base of the pavilion was originally a stone ladder. It was excavated by the villagers for the convenience of paper washing. Later, a brigade commander stationed here found it beguiling and built a pavilion on the site. Whether now or in the ancient past, whether literati or military attaches, people always appreciate its appeal. A few years ago, my classmate Ayu took me to visit Jinci Temple, but I did not pay attention to this wonderful place, so I highly recommend it this time.

2016.06.09 之上午
晋祠，难老泉，鱼沼亭

1-039 无锡，寄畅园七星桥（2019.03.22）

锦汇漪上七星桥后，鹤步滩与八音涧，未山先麓，借远景惠山。

1-039 Seven Star Bridge (*qixingqiao*) at the Jichang Garden, Wuxi (22 March 2019)

Hebutan Bank and Bayinjian Stream behind the Seven Star Bridge of the Jinhuiyi Pool, the foot of the mountain, borrowing the distant scenery of Huishan.

1-040 寄畅园，凤谷行窝园庭（2019.03.22）
秉礼堂临潭池，树石，水榭，碑廊与洞门。

1-040 Fenggu Xingwo Courtyard in the Jichang Garden, Wuxi (22 March 2019)

Bingli Hall faces a deep pool, stones and trees, waterside pavilion, stele corridor and cave door.

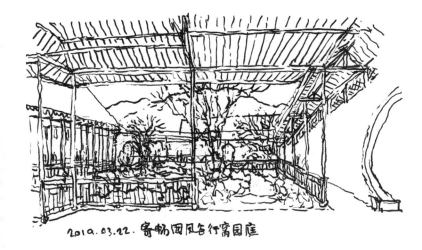

2019.03.22. 寄畅园凤谷行窝园庭

1-041 苏州，虎丘千人石 （2019.03.22）

假水浮石，剑池高桥，结合天然山石、地形造化的"造园"。
大小，高下，狭阔，真假。

1-041 Thousand People Rock (*qianrenshi*) at Tiger Hill (*huqiu*),
Suzhou (22 March 2019)

The fake water, pumice, sword-shaped (*Jianchi*), with its high
bridge, natural rocks and existing topography, were all incorporated
into a garden, embodying the rules of big and small, up and down,
narrow and wide, true and false.

2019.08.22
虎丘千人石

1-042 虎丘，剑池（2019.03.22）
　　层层树木，组构成深远之景。

1-042 Tiger Hill, Jian Pond (22 March 2019)
　　Many layers of trees form a view of deep distance.

虎丘
劍池

2019.03.22
尸尸88村廿
圆南岭公果路之景

1-043 虎丘，剑池旁"第三泉"之崖壁"微泉"（2019.03.23）

自然山石，细泉，砌石，泉道，书法石刻，浑然天成，格外动人。

1-043 Tiger Hill, the third spring - micro spring on the cliff side near Jian Pond (23 March 2019)

The natural rock, fine spring, stone masonry, spring channel, calligraphy and stone carving, like nature itself, are especially moving.

2019.03.23 虎口"第三泉"之飞泉崖壁

1-044 拙政园，由倒影楼南望起伏长廊、与谁同坐轩和廊后山上的宜两亭（2019.03.22）

童寯先生在《江南园林志》中以拙政园的空间塑造和体验过程为例，概述了造园有"三境界"：疏密得宜，曲折尽致，眼前有景。

1-044 The Humble Administrator's Garden, looking south from the Daoying Building to the undulating long corridor, Yushuitongzuo Pavilion and Yiliang Pavilion on the back of the mountain (22 March 2019)

In the Notes on Jiangnan Gardens (*Jiangnan yuanlin zhi*), Tong Jun used the space-shaping experience and process of the Humble Administrator's Garden as examples, and summarised the "three realms of gardening", including appropriate sparseness and density, twists and turns and dynamic scenery.

2019.03.22 拙政院与谁同坐轩

1-045 拙政园，由宜两亭俯望起伏长廊、与谁同坐轩和倒影楼（2019.03.23）

1-045 The Humble Administrator's Garden, overlooking the undulating long corridor, Yushuitongzuo Pavilion and the Daoying Building from Yiliang Pavilion (23 March 2019)

2019.03.23 拙政院宜两亭

1-046 拙政园，梧竹幽居 (2019.06.02)

"爽借清风明借月，动观流水静观山。"（梧竹幽居内对联）

鲁安东教授的文章《迷失翻译间：现代话语中的中国园林》是非常有启发的：游 vs 运动，景 vs 观，处 vs 空间，非正式性、沉浸性、想象性 vs 运动性、场地性、视觉性。特别是关于两个亭子的案例实在精彩，拙政园中的梧竹幽居曾带给我困惑，对"刘园"（留园）濠濮亭的故事则从未如此留意过，今天一并解惑了（2017.07.30）。

1-046 The Humble Administrator's Garden, Wuzhu youju Pavilion (2 June 2019)

Shuang jie qingfeng ming jie yue, dong guan liushui jing guan shan. (poetic couplet of Wuzhu youju Pavilion, meaning that borrowing the breeze makes one feel more refreshed, borrowing the brighter moon makes the light much clearer; watching the flowing water when you are moving, and watching the stable mountains when you are quiet)

The article *Lost in Translation: Chinese Garden in Modern Discourse* written by Lu Andong is very inspiring, mentioning the following points: wandering versus movement, scenery versus viewing, place versus space, informality, immersion, imagination versus movement, site, and visibility. What is particularly interesting is the case of the two pavilions. Wuzhu youju Pavilion in the Humble Administrator's Garden confused me before, and I had never noticed the story of Haopu Pavilion at the Liu Garden as much. Today, the article dispels both of my above doubts (30 July 2017).

远借（清风明月）
动观流水静观山
2019.06.02 拙政园 格竹鲤居

1-047 留园，濠濮亭（掬月亭）（2019.04.24）

　　与董老再次同游留园。仔细分辨了鲁安东研究发现的"五峰仙馆背后三岔立交"及"掬月 - 濠濮亭"变迁疑案；晚在平江路河边吃酒闲聊，看流水落花；返程高铁上无事，摹写刘懋功的《寒碧山庄图》（1857 年，源自安东兄讲座幻灯片），与园中景物回想参照了一下。

1-047 Haopu Pavilion (Juyue Pavilion) at the Liu Garden (24 April 2019)

　　I visited the Liu Garden with Dong Yugan again, carefully distinguishing the suspicious cases of the three interchanges behind the Wufengxianguan Pavilion and the change of the pavilion name from Juyue Pavilion to Haopu Pavilion that were studied by Lu Andong. In the evening, we drank wine and chatted on the riverside by Pingjiang Road, where we saw flowers falling into the river. I had nothing to do on the high-speed train back but to imitate Liu Maogong's Hanbi Mountain Villa (*Hanbi shanzhuang tu*, 1857, an image taken from Andong's lecture), referring to the scenery seen in the garden.

2019.04.24. 留园 濠濮亭(拙政园)

1-048 杭州，西湖，涌金桥（2019.04.20）
长桥行高低，湖水藏大小，群山望近远。

1-048 Yongjin Bridge, the West Lake, Hangzhou (20 April 2019)

The long bridge for walking is both high and low, the lake is divided into both big and small parts by the bridge, and the mountains are both near and far.

2019.04.20 杭州 西湖 海盏桥

1-049 杭州，灵隐飞来峰造像（之一）（2019.04.22）

青林洞。状如虎口，有前院、中庭、内院。佛像居于岩壁，或大或小，或群或单，或藏或显，或深或浅，与游人相映。树木葱郁，生长于岩崖之上和之间。

1-049 Statues on Feilai Peak at Lingyin Temple, Hangzhou (1) (22 April 2019)

The Qinglin Cave is shaped like a tiger's mouth, with front, central and inner courtyards. Buddha statues inhabit the rock wall, large and small, in groups or single, hidden or visible, deep or shallow, interacting with visitors. The trees are lush and grow on top of the cliffs or between them.

2019.04.22
杭州灵隐飞来峰青林洞

1-050 杭州，灵隐飞来峰造像（之二）（2019.04.22）

观音洞，"一线天"。数尊单一佛像、龛状空间自由居于山体，有临半悬空崖壁之上者，其下有石级通冷泉溪流之上，近有桥亭。

1-050 Statues on Feilai Peak at Lingyin Temple, Hangzhou (2) (22 April 2019)

There is a view of a thin strip of sky inside the Guanyin Cave. Several single Buddha statues in niches are freely carved into the mountains, some of which are semi-suspended on the cliffs. Below them is a stone ladder reaching down to the cold spring stream near the bridge pavilion.

2019.04.22　　秋林吴隐石栗峰　　"聊世夫"

1-051 杭州，灵隐飞来峰造像（之三）（2019.04.22）

冷泉溪。弥勒等佛像沿溪边丛岩高低分布，有游人沿石阶渐上渐隐于山中，山上满布树林：水—石—树—人—像—阶。

1-051 Statues on Feilai Peak at Lingyin Temple, Hangzhou (3) (22 April 2019)

It is the Cold Spring Creek. The statues of Maitreya and other Buddhas are distributed along the cliffs beside the creek. Tourists walking on the stone steps are gradually hidden by the mountains covered with trees: creek, stones, trees, people, statues, steps.

2019.04.22 冷泉溪　　　杭州灵隐飞来峰造像

1-052 杭州，西泠印社，文泉与闲泉 （2019.04.23）

一处极有人文特色的山地小园，两泉池缩窄相通，高低平台，水边步道，阶梯石桥，隧洞经塔，佛龛塑像，竹林树木，大小文字刻于岸边石壁。

1-052 Hangzhou, Xiling Seal Engravers' Society, Wen Spring and Xian Spring (23 April 2019)

This is a small mountain garden with extremely humanist features: two spring pools that become narrow and connected, high and low platforms, waterside trails, stepping stone bridges, buddhist towers and the tunnels, Buddhist statues in niches, bamboo and trees, large and small characters engraved on the shore's stone wall.

1-053 杭州，西泠印社，二泉与小龙泓洞、华严经塔（之二）
（2019.04.23）

　　仿佛高浮雕石刻印章放大而成的微缩小园，却有深远无尽之妙。

1-053 Hangzhou, Xiling Seal Engravers' Society, Er Spring,
Xiaolonghong Cave and Huayan Tower (2) (23 April 2019)
　　The micro-garden looks like an enlarged stone carving seal with
high relief and has far-reaching, endless excellence.

2019.04.23 杭州西泠印社 二泉 + 华严经塔 小龙泓洞

1-054 杭州，西泠印社，四照阁旁（2019.04.23）。
湖山一览。

1-054 Hangzhou, Xiling Seal Engravers' Society, next to Sizhao Pavilion (23 April 2019)

　　Overlooking lakes and mountains.

2019.04.23
杭州西湖卸社湖山一楼

1-055 杭州，虎跑泉，泊云桥＋日月池＋含晖亭 (2019.04.24)

含晖亭实为虎跑景区真正入口，与日月池、泊云桥一体结合，又桥面结合坡道台阶，水池汇引泉水，内外方向不同而景致皆佳，为巧妙独特的营造。

1-055 Boyun Bridge, Riyue Pond and Hanhui Pavilion, Hupao Spring, Hangzhou (24 April 2019)

Hanhui Pavilion, as the real entrance to the Hupao Scenic Area, is integrated with Riyue Pond and Boyun Bridge, whose deck combines a ramp with steps. The spring flows into the pond. From the inside and outside there are excellent views of the scenery in different directions. What a wonderful design and construction!

2019.04.24. 杭州西湖

1-056 杭州，虎跑泉，品泉阁（2019.04.24）

于品泉阁饮尝用虎跑泉水冲泡的西湖龙井"双绝茶"，自茶座望罗汉堂与碑廊，两者之间对角自然相接，透漏出后方山林石阶，空间幽然深远。自然而然，容纳偶然。

1-056 Pinquan Pavilion, Hupao Spring, Hangzhou (24 April 2019)

In Pinquan Pavilion, I taste the West Lake Longjing tea brewed with Hupao spring water, while looking at the Luohan Hall and stele corridor that are connected diagonally. Behind them are mountains, trees and stone steps, with quiet and far-reaching spaces. This natural space can accommodate all kinds of events.

2019.06.24. 杭州 虎跑泉
四览而绘。整体偶然

1-057 扬州，何园（寄啸山庄），片石山房平面及轴测草图（2015.02.25）

何园是扬州私园"压轴之作"，"天下第一廊"——长1500米的"复道"回廊环绕楼阁、假山，又以二层"串楼"和回廊连成一片，左右分院，高低勾搭，衔山环水，登堂入室；"天下第一山"——园中园片石山房之假山，石涛"人间孤本"。

1-057 He Garden (Jixiao Villa), Yangzhou. Plan and axonometric sketch of Pianshi shanfang Garden (25 February 2015)

He Garden is the masterpiece of the Yangzhou private garden. Its 1,500-metre long double-corridor surrounds the pavilion and the rockery, connecting the cloister by a two-storey building. The corridor divides the courtyard into two parts. It joins the upper and lower, links up the hill, surrounds the water and leads to the rooms. The "first mountain in the world", the rockery of Pianshi shanfang, a garden in the garden, is the only work existing today created by the Qing Dynasty painter Shi Tao.

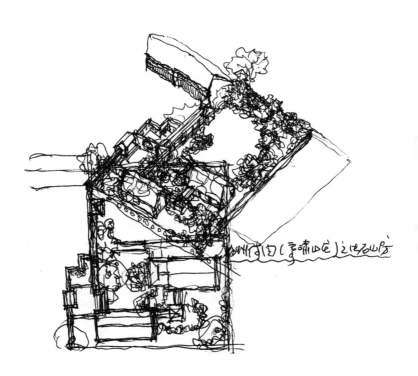

苏州留园（寒碧山庄）之揖峰山房

1-058 何园（寄啸山庄），片石山房，由门厅望水榭及假山（2015.02.25）

以湖石紧贴墙壁堆叠假山，下屋上山，山上寒梅与罗汉松，合画理"左急右缓，切莫两翼""山欲动而势长"。整个山体均为小石叠砌而成，故名"片石山房"，石块拼镶技巧极为精妙，拼接之处有自然之势而无斧凿之痕，气势、形状、虚实处理与石涛画相符。山体环抱水池，池南水榭，遥对假山主峰，崖壑流云、茫茫烟水。门厅"注雨观瀑"，书屋、棋室、琴台、竹石图，"琴棋书画"合为一体。

2015-02-25 扬州何园 片石山房

1-058 He Garden (Jixiao Villa), Pianshi shanfang Garden. Looking at the waterside pavilion and rockery from the entrance hall (25 January 2015)

The rockery is stacked with lake stones placed close to the wall. In the lower part there is a house, and in the upper part there is a mountain where plum trees and yew plum pines grow. Such a situation corresponds to painting theories of the steep left and the gentle right with differentiation on both sides and long mountains with dynamic shapes. The whole mountain is made up of small stones, so the name is Pianshi shanfang. The stone inlaying technique is exquisite. The splicing point has a natural appearance without axe and chisel marks. The rockery's momentum, shape, void-space and volume-form are consistent with Shi Tao's painting. The mountain surrounds a pool and a waterside pavilion is located to the south of the pool facing the main peak of the rockery. It forms an image consisting of a cliff, water, clouds and fog. The entrance hall, study, chess room, musical instrument table and bamboo stone painting are integrated.

1-059 南京，瞻园（红枫院）（2015.07.22）
"柱林"，层叠的空间，高下。

1-059 Hongfeng Courtyard, Zhan Garden, Nanjing (22 July 2015)
  "A forest of columns", juxtaposed spaces, high and low.

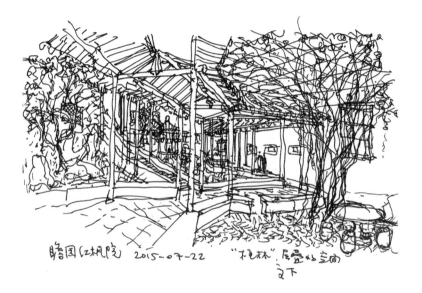

瞻园（江南院） 2015-07-22 "桓林" 屋盖加宝顶
之下

1-060 冯纪忠，上海松江，方塔园（之一）（2013.06.05）
俯瞰堑道，错动，放缩，大小，高下。

1-060 The Garden of the Square Pagoda designed by Feng Jizhong, Songjiang, Shanghai (1) (5 June 2013)

Overlooking the trench corridor, the space of dislocation, enlarging and shrinking, big and small, high and low.

1-061
冯纪忠，上海松江，方塔园（之二）（2013.06.05）
堑道倾斜如山石，上植密林，树石空间。

1-061 The Garden of the Square Pagoda designed by Feng Jizhong,
Songjiang, Shanghai (2) (5 June 2013)

The inclined trench corridor looks like rocks, with trees growing
on the top. It is a space composed of trees and stones.

1-062 冯纪忠，上海松江，方塔园（之三）（2013.06.05）

堑道出口，曲折尽致。

1-062 The Garden of the Square Pagoda designed by Feng Jizhong, Songjiang, Shanghai (3) (5 June 2013)

The exit to the trench corridor, with twists and turns.

1-063 冯纪忠，上海松江，方塔园（之四）（2013.06.05）
曲墙密林后的何陋轩。

1-063 The Garden of the Square Pagoda designed by Feng Jizhong,
Songjiang, Shanghai (4) (5 June 2013)
The Helouxuan tea house behind the curved walls and trees.

1-064 冯纪忠，上海松江，方塔园（之五）（2013.06.05）

何陋轩座内外望，金黄的屋顶，压低压暗的檐口，外面明亮的绿植和水面。

1-064 The Garden of the Square Pagoda designed by Feng Jizhong, Songjiang, Shanghai (5) (5 June 2013)

Looking out from the Helouxuan tea house, there are golden roof, low and dark cornice, bright green plants and water.

1-065 北京，真觉寺，五塔＆双杏（2017.01.31）
"五"对"双"，"塔"对"杏"，对仗。

1-065 Five towers and double apricots in Zhenjue Temple, Beijing (31 January 2017)

Five versus double, tower versus apricot, an antithesis.

2019. 01. 31. 真觉寺
金刚宝座

1-066, 067, 068, 069

树石四则（2018.01.07，2019.05.24，2019.06.03）

中国赏玩文化中特有的盆景，可被视为一种极度缩微的园林，亦或真实山水的模型，其底盘则可视作园林的边界甚至围墙。其树，其石，其水（底盘的空白之地），甚至小小亭台楼阁，完全如园林一般，参照山水画小品的空间构图，只不过除了有些以石为主景的盆景外，树木通常被强调为主景。

1-066, 067, 068, 069 Four sketches of stones and trees (7 January 2018, 24 May 2019, 3 June 2019)

The unique bonsai garden appreciated in Chinese culture can be regarded as an extremely miniature garden, or a model of a real landscape. Its base plate can be viewed as the boundary of the garden or even the bounding wall. Its trees, stones, water (the blank space of the base plate), and even the small pavilions are completely similar to full scale gardens, with reference to the spatial composition of landscape paintings. Usually trees are emphasised as the main scene, except for some bonsais with stone-themed landscapes.

2018.01.07.　　　深谷春鸣

山林雅趣

2018.01.08.

最为重要的是，观赏者需要在想象中缩小为一个小小的人，并将自己化身入这"拳石勺水"的风景中，去"卧游"，如宗炳在《画山水序》中所言者。与董豫赣兄再游留园时，在盆景园中发现一椭圆盆景卧石与我们的乐高2号极相似，但又与乐高2号的来源——《素园石谱》中的"永州石"决然不同，有趣。

The most important thing is that the viewer must imagine himself in miniature, transforming himself to fit into the tiny scenery (*quanshi shaoshui*, or fist-sized stones and a spoonful of water) and imaging travelling as he would in a real landscape (*wo you*), as Zong Bing argued in the Preface of Landscape Painting (*Hua shanshui xu*). Interestingly, visiting the Liu Garden again with Dong Yugan, I find that an ellipsoidal stone in the bonsai garden is very similar to our Lego 2 project, but different from the Yongzhou Stone of the *Suyuan shipu*, the origin of the Lego 2 project.

树石一20 / 2019.06.03   深谷春鸣

2019.5.24. 留园. 像"朵高一号"的山石盆景

1-070 (1-5) 《乌有园袖峰石谱》 (2016.09.03)

周末会议无聊犯困中，不觉成此"石谱"。边画边羡慕王欣兄台：袖藏这么多超绝解闷之尤物啊（最后的"环形山"除外）。袖石巨峰，在人的观想中，小大转换。

1-070 (1-5) *Wuyouyuan xiufeng shipu* (3 September 2016)

The weekend meeting was so boring that I felt sleepy, and I did not realise that I had been making a booklet with drawings of stones. While painting, I envied Wang Xin, who has hidden so many interesting stones (except the ring-shaped stone). In people's imagination, the small stone and the giant peak can be converted.

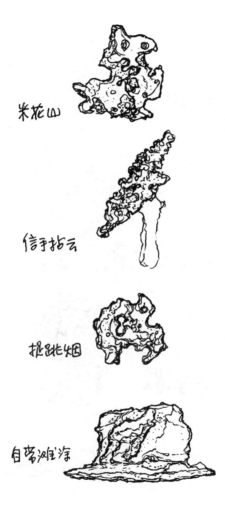

米花山

信手拈云

挑跳烟

自客滩净

风砺山殿

横云搁挡

海噬博山

匀波裹一眠

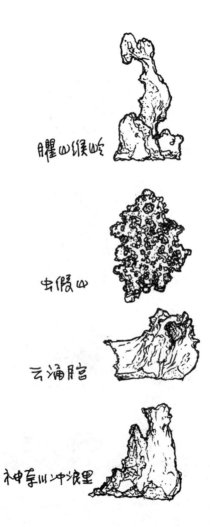

瞿山猴岭

出假山

云涌脂

神奈川冲浪里

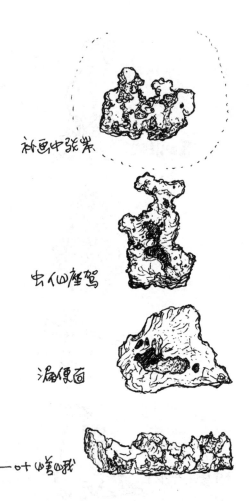

补画中张某

出仙崖驾

海便面

一叶山笑山我

153

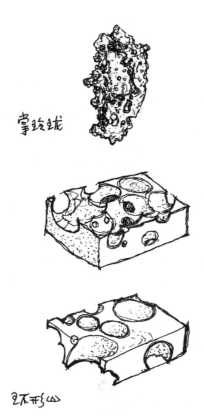

掌玲珑

卫不形山

1-071 日本金泽市，桥之作：兼六园虹桥（之一）(2016.12.17)

1822 年的桥作。六胜：宏大，幽邃，人力，苍古，水泉，眺望。

1-071 The rainbow-shaped bridge in the Kenroku-en Garden, Kanazawa City, Japan (17 December 2016)

The bridge was made in 1822. The six features of the garden are grandness, depth and serenity, manpower, historical vicissitude, a water spring and an overlooking stance.

2016.12.17. 日本. 金泽 兼六隆园. 虹桥　1822
大胜：宏大·照遊·人力·苍古·水泉·眺望

1-072 日本金泽市, 桥之作: 兼六园虹桥(之二) (2016.12.18)

　由近及远: 雪后铺上金黄色草垫的弧形板拱桥面, 石灯笼, 松树, 水岸, 架空的水边房屋, 码头。

1-072 The rainbow-shaped bridge in the Kenroku-en Garden, Kanazawa City, Japan (18 December 2016)

　From the near to the distant are the curved arch bridge deck covered by golden straw mats after the snow, stone lanterns, pine trees, a waterfront, overhead waterfront houses, and a dock.

2016.12.18. 兼六园·梓江作

1-073 日本岐阜县大野郡，桥之作：通往白川乡的悬索桥

（2016.12.07）

1-073 The suspended bridge to Shirakawago, Ono-gun, Gifu
Prefecture, Japan (7 December 2016)

2·16.12.07 日本 白川乡   江户后期

1-074,075,076 不食人间烟火的日本禅寺园庭／

茶室——京都，四寺四庭：南禅寺金地院鹤龟之庭，高台寺园庭，

大德寺聚光院园庭，仁和寺园庭（2016.12.19）

　　树、石、沙／池，与放大的盆景类似，但与中国的盆景不同，
更强调面向而观的视觉效果，并不欢迎人的进入（包括想象中的
进入）。

2016.12.09. 京都高台寺园庭

1-074, 075, 076 Otherworldly Japanese Zen temple garden/tea room in Kyoto, the temples and four gardens: Nanzen-ji Temple's Konchi-in Crane and Turtle Garden, K ō dai-ji Temple Garden, Daitoku-ji Temple's Juk ō -in Garden, Ninna-ji Temple Garden (19 December 2016)

Trees, rocks, sand/pools, similar to magnified bonsai, but unlike Chinese bonsai, they emphasise the visual effects of facing and viewing, rather than welcoming people's entry (including imaginary entry).

2016.12.19 京都 南禅寺 金地院 鶴亀之庭 亀石, 小堀遠州,
1629年

2016.12.19 聚光院 园庭，千利休（相伝），倍果树林

1-077 形式之美：京都，东福寺开山堂（2016.12.20）

不对称构图的强调。

1-077 The beauty of form: T ō fuku-ji Temple's Kaisando Hall, Kyoto (20 December 2016)

The emphasis is on asymmetric composition.

2016.12.20 东都东福寺 开山堂

1-078 形式之美：京都，东福寺方丈庭入口（库堂）（2016.12.20）

　　大山墙面＋披檐＋卷蓬入口——其实是变形了的歇山顶和"老虎窗"。

1-078 The beauty of form: entrance to the Honbo Garden (Kutang) of T ō fuku-ji Temple, Kyoto (20 December 2016)

　　The large gables, apron eave and canopy entrance are, in reality, a deformed gable and hip roof and dormer.

2016.12.20 京都东福寺方丈庭入口（库里）

行／旅（山水与旅行）

　　园林与山水画对于其中的观想者都是"卧游"，即南朝宋的宗炳"凡所游历，皆图于壁，坐卧向之"，"澄怀观道，卧以游之"，而宗炳之卧游，是本来"好山水，爱远游"，但"老疾俱至，名山恐难遍睹"的无奈之举。唯有亲身游览过真实的名山巨川、苍峰流瀑，才能理解和进入那些缩微于山水画和园林之中的山水空间与意境。

　　自然是人们摹写、学习与介入的对象，是人文艺术的不尽源泉。在自然中旅行、游憩，是一种人与时间和空间的结合，与精神世界的心灵体验和人生感悟密切相连。时间和空间的结合会产生特殊的诱惑——被赋予时间感的空间和物体能够激发生命的体验。这种时间感的第一种是由人在空间中的运动而产生；另一种时间感是由人面对空间的深远层次而产生，就如"山外有山"，那样层层不断地向深处向远处延伸；最后一种时间感则是由"硬生生的"时间和生命自身而产生，十年、百年、千年，生涩、成熟、衰老、死亡。空间感—时间感—生命感彼此链接作用于人的内心世界，导致诗意的产生。

　　只有在山中行走，才能真正理解山——如果不用手脚、眼睛和内心亲身体验真正的自然山水，怎么能产生与山水关联的想象和营造呢？于是我曾经实地游览长江三峡、黄山、敦煌莫高窟、洛阳龙门石窟、安吉梅溪等，在行旅中体验、对照、想象、理解那些先人在山水绘图与造园中所呈现的意象和境界。

Walking
(Landscape and traveling)

Appreciating gardens and landscape paintings is another way of travelling, or *wo you*. That is, as Zong Bing of the Song dynasty during the Northern and Southern dynasties argued, all of the scenery observed during travel can be drawn on a wall so that one can sit down and view it. Zong Bing was a travel enthusiast. However, his engagement in *wo you* was because of aging and illness, and unable to visit all of the famous mountains. Only by experiencing natural landscapes can one truly understand the artistic conception that is embodied in landscape paintings and gardens.

Nature, for human beings, is an object of depicting, learning and intervention, and an inexhaustible inspiration for the arts and humanities. Travelling and recreation in nature is a combination of people, time and space. It is closely related to spiritual experiences and reflections on life. The combination of time and space creates a special temptation. Spaces and objects with a sense of time can stimulate life experience. The first sense of time is produced by bodily movement within space. The second sense of time is produced when people face multi-layered spaces, just like mountains beyond mountains, which are constantly extending to the far-reaching depths. The last sense of time is produced by time and life itself, ten years, centuries and millennia; being immature, mature, aging and dead. The sense of space, time and life that are linked to each other impact people's inner worlds, leading to the creation of poetry.

Only by walking in the mountains can one understand them. If you do not experience the natural landscape with your hands, feet, eyes and heart, how can you create the images and objects associated with mountains and rivers? I have visited places like Yangtze River, Three Gorges, Huangshan Mountain, Dunhuang Mogao Grottoes, Luoyang Longmen Grottoes and Anji Meixi to experience, contrast, imagine and understand the images and realms presented in the ancestors' landscape painting and gardening.

2-001 黄山记游：黄山玉屏峰迎客松（2015.09.16）

空间深远。

2-001 Notes on the Huangshan tour: Greeting Pine on Huangshan Yuping Peak (16 September 2015)

The space is far-reaching.

2015.09.16. 黄山
玉屏索道
KP 寺太乃

2-002 黄山记游：黄山天都峰 （2015.09.17）

石庭深深。

2-002 Notes on the Huangshan tour: Huangshan Tiandu Peak (17
September 2015)

The stone courtyard is deep.

2015 0917

黄山 天都峰

2-003 黄山记游：黄山玉屏峰（2015.09.17）

人（工）游憩于自然，高远。

2-003 Notes on the Huangshan tour: Huangshan Yuping Peak (17 September 2015)

People wandering and recreating in nature; high distance.

黄山
（天都峰、
玉屏峰）

2015.09.17

2-004 黄山记游：黄山莲花峰（之一）（2015.09.17）

某洞天内观。

2-004 Notes on the Huangshan tour: Huangshan Lotus Peak (1) (17 September 2015)

Looking outside from the interior of a fairyland.

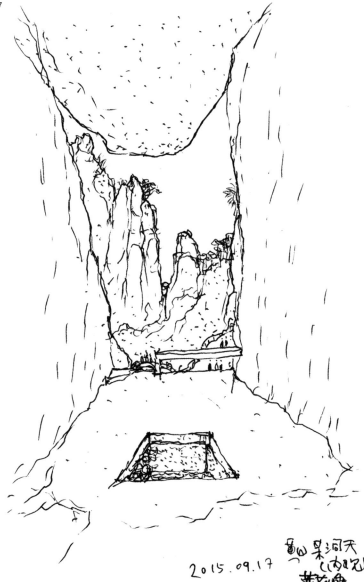

黄山 果洞天
(内观)
莲花峰

2015.09.17

2-005 黄山记游：黄山莲花峰（之二）（2015.09.17）

某洞天外观。

2-005 Notes on the Huangshan tour: Huangshan Lotus Peak (2) (17 September 2015)

The exterior view of a fairyland.

莲花峰

2015 09 13 黄山莲花峰天

(外双沟)

2-006 黄山记游：黄山鳌鱼峰（天海）（2015.09.16）

前山看后山，如门，层层打开。

2-006 Notes on the Huangshan tour: Huangshan Aoyu Peak
(Tianhai) (16 September 2015)

Standing in front of the mountain and looking at the mountains
beyond, one can observe scenery that is similar to many layers of
doors opening.

2015.9.16 黄山 （北海）
鳌鱼峰

2-007 黄山记游：黄山西海大峡谷（2015.09.18）

深远峰景，索桥，火车。

2-007 Notes on the Huangshan tour: Huangshan Xihai Grand Canyon (18 September 2015)

Far-reaching peak views, cable bridges, trains.

2015 09 18 黄山 西海大峡谷

2-008 黄山记游：黄山西海大峡谷 （2015.09.18）

人在山林之海中忽隐忽现，蹬道与平台。

2-008 Notes on the Huangshan tour: Huangshan Xihai Grand Canyon (18 September 2015)

　　Images of people flickering in the mountains and forests; steps and platforms.

2015.09.18 黄山 西海大峡谷
（10.03 补画）

2-009 黄山记游：黄山始信峰（之一）（2015.09.18）

深远。

2-009 Notes on the Huangshan tour: Huangshan Shixin Peak (1) (18 September 2015)

Deep distance.

2015 04 18 黄山始信四峰

2-010 黄山记游：黄山始信峰（之二）（2015.09.18）

高远＋平远。

2-010 Notes on the Huangshan tour: Huangshan Shixin Peak (2) (18 September 2015)

High distance and level distance.

2015·09·18 黄山始信峰

2-011 黄山记游：宏村（2015.09.16）

宏村街巷，空间曲折，连续变化，而山沉默不动，永为衬景。

2-011 Notes on the Huangshan tour: Hong Village (16 September 2015)

Hong Village streets have twists and turns and continuous changes against the backdrop of the silent mountains.

2015.09.16. 宏村

2-012 黄山记游：绩溪棋盘村 （2015.09.19）

一个棋盘格局的北方村落，坐落于徽州山水之间。

2-012 Notes on the Huangshan tour: Qipan Village in Jixi (19 September 2015)

A village structured with a northern-style checkerboard pattern is located between the mountains and rivers of Huizhou.

2015·09·19. 清溪石家村（楂鹽村）

2-013 长江记游：重庆（之一）（2014.10.13）

城市拼贴。

2-013 Notes on Yangtze River travel: Chongqing (1) (13 October 2014)

City collage.

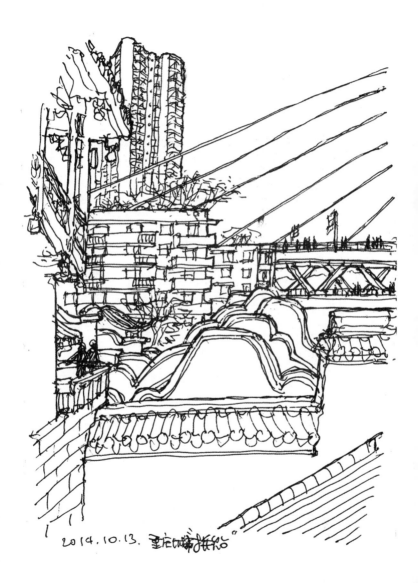

2014.10.13. 王宏建

2-014 长江记游：重庆（之二）（2014.10.01）

山地城市街道界面对江景、山景、城景所形成的"过白"。

2-014 Notes on Yangtze River travel: Chongqing (2) (1 October 2014)

The framed scene (*guobai*) formed by the interfaces of urban streets with the river, mountain and city.

2-015 长江记游：重庆（之三）（2014.10.14）

江岸，城市，双桥。

2-015 Notes on Yangtze River travel: Chongqing (3) (14 October 2014)

River bank, the city, double bridges.

2-016 长江记游：重庆，忠县，石宝寨 （2014.10.14）
山崖之上的村寨。

2-016 Notes on Yangtze River travel: Shibaozhai Village, Zhong County, Chongqing (14 October 2014)

A village above the cliff.

2016.10.10. 重庆·忠县·石宝寨

2-017 长江记游：白帝城（2014.10.14）

三峡蓄水后，白帝城从半岛变为真正的岛，由桥与主山相连。

2-017 Notes on Yangtze River travel: Baidi City (14 October 2014)

Through the storage of water in the Three Gorges Reservoir, Baidi City has been transformed from a peninsula into a real island, connected to the main mountain by a bridge.

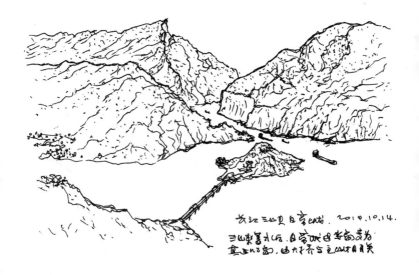

长江 三峡夔门 白帝城站. 2010.10.14.

三峡最早水库. 白帝城是白帝庙为
在 山上的岛, 由大桥与三峡相连可上白天

2-018 长江记游：夔门（之一）（2014.10.14）
由奉节白帝城俯望。

2-018 Notes on Yangtze River travel: Kui Men (1) (14 October 2014)
Looking down from Baidi City, Fengjie.

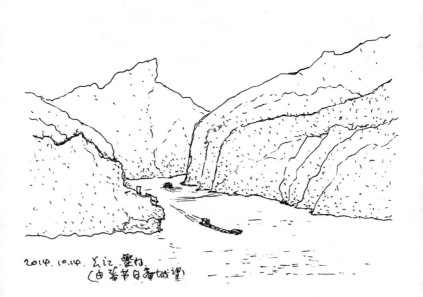

2-019 长江记游：夔门（之二）（2014.10.14）

相信是江水在急转弯处的力量，历经漫长时间，"切削"出如此陡峭垂直的崖壁。

2-019 Notes on Yangtze River travel: Kui Men (2) (14 October 2014)

It is believed that the strength of the river at the sharp turn, after a long period, cut such a steep vertical cliff.

2014.10.14. 长江三峡之 夔门

2-020 长江记游：夔门（之三）（2014.10.14）

巨大如门的礁石。

2-020 Notes on Yangtze River travel: Kui Men (3) (14 October 2014)

A huge reef like a gate.

2014. 10. 14. 한라골프관광

2-021 长江记游：夔门（之四）（2014.10.14）
即将入"门"。

2-021 Notes on Yangtze River travel: Kui Men (4) (14 October 2014)
Approaching the "gate".

2014.10.14.

2-022 长江记游：夔门（之五）（2014.10.14）
进入"门"内。

2-022 Notes on Yangtze River travel: Kui Men (5) (14 October 2014)
   Entering the "gate".

2014.10.14. 长江三峡夔门

2-023 长江记游：夔门（之六）（2014.10.14）
出"门"后的开阔江面处。

2-023 Notes on Yangtze River travel: Kui Men (6) (14 October 2014)
The river becomes wide out of the "gate".

2014.10.14. 长江记游 开阔江面处

2-024 长江记游: *存在于自然之中的几何（之一）*（2014.10.14）

30° 三角形的山。

2-024 Notes on Yangtze River travel: geometry existing in nature (1)
(14 October 2014)

A thirty-degree triangular mountain.

2014.10.14. 长江三峡。三十度三角形的山

2-025 长江记游: 存在于自然之中的几何（之二）(2014.10.14)

45°三角形的山。

2-025 Notes on Yangtze River travel: geometry in nature (2) (14 October 2014)

A forty-five-degree triangular mountain.

2-026 长江记游: 存在于自然之中的几何(之三)(2014.10.14)
　　等边三角形的山。

2-026 Notes on Yangtze River travel: geometry in nature (3) (14 October 2014)
　　A mountain shaped like an equilateral triangle.

2014.10.14. 三日坐. 阿坝州三角形的神山

2-027 长江记游: 存在于自然之中的几何(之四)(2014.10.14)
很多等边三角形的山。

2-027 Notes on Yangtze River travel: geometry in nature (4) (14
October 2014)

Many mountains shaped like equilateral triangles.

像刀削过三角形的山， 2014.10.14. 长江三峡头

2-028 长江记游: 存在于自然之中的几何(之五)(2014.10.14)

很多三角形的山。

2-028 Notes on Yangtze River travel: geometry in nature (5) (14 October 2014)

Many mountains in triangular shape.

很多之前形的上山   2014.10.14. 与已二雄之

2-029 长江记游：居在江岸（之一）（2014.10.14）

江边建筑。

2-029 Notes on Yangtze River travel: living on the river bank (1) (14 October 2014)

Riverside architecture.

2-030 长江记游：居在江岸（之二）（2014.10.14）

江岸人家。

2-030 Notes on Yangtze River travel: living on the river bank (2) (14 October 2014)

Riverside homes.

2-031 长江记游：居在江岸（之三）（2014.10.14）
由江边蜿蜒爬升到半山的建筑。

2-031 Notes on Yangtze River travel: living on the river bank (3) (14 October 2014)

Riverside buildings extending halfway up the mountain.

2014.10.14. 毛泽东三峡水

2-032 长江记游：居在江岸（之四）（2014.10.14）

宜昌三斗坪，山城＋江城。

2-032 Notes on Yangtze River travel: living on the river bank (4) (14 October 2014)

Sandouping Town in the city of Yichang, a mountain city and a river city.

2014.10.14. 宜昌三江大桥 山包的 + 记城

2-033 长江三峡工程模型（2014.10.14）

三峡工程——世界上最复杂的双线五级船闸，目前世界上总水头最高、连续级数最多的大型船闸。

2-033 Model of the Yangtze Three Gorges Project (14 October 2014)

The Three Gorges Project, the world's most complex two-line five-level ship lock, currently has the world's tallest hydraulic head and largest number of continuous ship lock levels.

227

2-034 敦煌，鸣沙山 （2015.06.20)
沙山，绿棘。

2-034 Mingsha Hill, Dunhuang (20 June 2015)
Sand hills, green spines.

敦煌鸣沙山　2015－06－20

2-035 敦煌，鸣沙山，月牙泉（之一）（2015.06.20）
正面：沙山，月泉，楼阁。

2-035 Moon Spring, Mingsha Hill, Dunhuang (1) (20 June 2015)
Front: sand hill, Moon Spring, and a pavilion.

鸣沙山 月牙泉 2015
06
20

2-036 敦煌，鸣沙山，月牙泉（之二）（2015.06.20）
侧面：沙山，月泉，楼阁。

2-036 Moon Spring, Mingsha Hill, Dunhuang (2) (20 June 2015)
Side: sand hill, Moon Spring, and a pavilion.

·鸣沙山 月牙泉 2015 06 20

2-037 敦煌，鸣沙山，月牙泉（之三）（2015.06.23）

坐在巨大的沙山上俯望月牙泉和月泉阁：沙漠"抽象"背景中的山、水、花木、楼阁，人工之泉，幻境之感。

2-037 Moon Spring, Mingsha Hill, Dunhuang (3) (23 June 2015)

Sitting on a huge sand hill overlooking the Moon Spring and Moon Spring Pavilion, the mountains, water, flowers, trees, pavilions, man-made springs and sense of fantasy in the abstract background of the desert.

鸣沙山月牙泉
2015-06-23

2-038 敦煌，鸣沙山，月牙泉（之四）（2015.06.20）
月泉阁上望沙山。

2-038 Moon Spring, Mingsha Hill, Dunhuang (4) (20 June 2015)
Looking at the sand hill from Moon Spring Pavilion.

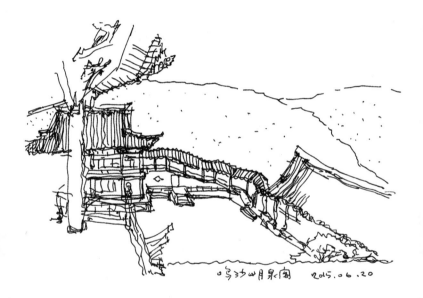

鸣沙山月泉阁　2015.06.20

2-039 敦煌，莫高窟（之一）（2015.06.21）
　　由对面的山坡遥望莫高窟。

2-039 Mogao Grottoes, Dunhuang (1) (21 June 2015)
　　Looking at the Mogao Grottoes from the opposite hillside.

敦煌莫高窟 2015 06 21

2-040 敦煌，莫高窟（之二）（2015.06.21）
莫高窟北区：仿佛孔洞沙山的断面。

2-040 Mogao Grottoes, Dunhuang (2) (21 June 2015)
North of Mogao Grottoes looks like a section of a sand hill with holes.

莫高窟北区 20150621

2-041 敦煌，莫高窟（之三）（2015.06.21）

莫高窟南区。入口前序：大桥—河道—小广场—白杨林—甬道—庭院—窟区。

2-041 Mogao Grottoes, Dunhuang (3) (21 June 2015)

South of Mogao Grottoes; entrance sequence: bridge, river, small square, poplar forest, paved path, courtyard and cave area.

莫高窟南区 2015 06 21

入口区：大桥孔河道、小火沟白杨林、南道、南院之窟区。
前序

2-042 敦煌，玉门关小方盘城（之一）（2015.06.22）

由自然和时光赋予其极简而复杂的形式和空间。

2-042 The small Fangpan Castle at Yumenguan, Dunhuang (1) (22 June 2015)

A minimalist and complex form and space created by nature and time.

玉门关小方城
2015.06.22

2-043 敦煌，玉门关小方盘城（之二）（2015.06.22）

高天旷沙，丘阜莽草，残城墟门，苍貌秘境。

2-043 The small Fangpan Castle at Yumenguan, Dunhuang (2) (22 June 2015)

High sky and empty desert, upland and wild grass, ruined gate and remnant city, pale appearance and mysterious surroundings.

玉门关小方盘城

2015-06-22

2-044 敦煌雅丹（之一）（2015.06.22）
造化抽象山水。

2-044 Dunhuang Yadan (1) (22 June 2015)
  Nature makes the abstract landscape.

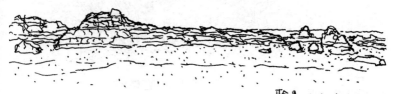

雅丹 2015 06 22

2-045 敦煌雅丹（之二）（2015.06.22）
平远山水。

2-045 Dunhuang Yadan (2) (22 June 2015)
Level distance landscape.

2-046 敦煌雅丹（之三）（2015.06.22）

深远山水。

2-046 Dunhuang Yadan (3) (22 June 2015)

Deep distance landscape.

雅丹 2015·06·22

2-047 敦煌雅丹（之四）（2015.06.22）
"沧海桑田"一。

2-047 Dunhuang Yadan (4) (22 June 2015)
Vicissitudes of time (1).

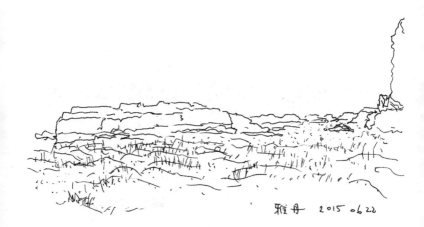

雅丹 2015 06 22

2-048 敦煌雅丹（之五）（2015.06.22）
"沧海桑田" 二。

2-048 Dunhuang Yadan (5) (22 June 2015)
   Vicissitudes of time (2).

雅丹 2015 0622

2-049 敦煌雅丹（之六）（2015.06.22）
"沧海桑田"三。

2-049 Dunhuang Yadan (6) (22 June 2015)
Vicissitudes of time (3).

2-050 敦煌雅丹（之七）（2015.06.22）

枯山水，大海样一。

2-050 Dunhuang Yadan (7) (22 June 2015)

Rock garden, imitating the surface of the sea (1).

雅丹：枯山水-大海样
2015 06 22

2-051 敦煌雅丹（之八）（2015.06.22）
枯山水，大海样二。

2-051 Dunhuang Yadan (8) (22 June 2015)
Rock garden, imitating the surface of the sea (2).

2-052 敦煌雅丹（之九） （2015.06.22）
枯山水，大海样三。

2-052 Dunhuang Yadan (9) (22 June 2015)
　　Rock garden, imitating the surface of the sea (3).

罗怀月　2015 06 22

2-053 踏访梅溪：行，望，居，游（之一）（2015.11.18）

绍兴新昌，金山村。

2-053 Visiting Meixi. Walking, viewing, living and travelling (1) (18 November 2015)

Jinshan Village, Xinchang County, Shaoxing.

2015.11.18. 金山村

2-054 踏访梅溪：行，望，居，游（之二）（2015.11.18)

绍兴新昌，梅坑盘山寺遗址。

2-054 Visiting Meixi. Walking, viewing, living and travelling (2) (18 November 2015)

Ruin of Panshan Temple in Meikeng Village, Xinchang County, Shaoxing.

2015.11.18. 梅坑盘山寺（遗址）

2-055 踏访梅溪：行，望，居，游（之三）（2015.11.18）
绍兴新昌，梅坑村。

2-055 Visiting Meixi. Walking, viewing, living and travelling (3) (18 November 2015)

Meikeng Village, Xinchang County, Shaoxing.

2-056 踏访梅溪：行，望，居，游（之四）（2015.11.18）
绍兴新昌，观音山与大坪头村。

2-056 Visiting Meixi. Walking, viewing, living and travelling (4) (18 November 2015)

    Guanyin Mountain and Dapingtou Village, Xinchang County, Shaoxing.

2-057 踏访梅溪: 行, 望, 居, 游（之五）（2015.11.18）
绍兴新昌, 内湾。

2-057 Visiting Meixi. Walking, viewing, living and travelling (5) (18 November 2015)

Neiwan, Xinchang County, Shaoxing.

2-058 踏访梅溪：行，望，居，游（之六）（2015.11.18）
绍兴新昌，胡卜村，七星峰。

2-058 Visiting Meixi. Walking, viewing, living and travelling (6) (18 November 2015)

Qixing Peak, Hubu Village, Xinchang County, Shaoxing.

七星峰

大坝

2015.11.18   胡卜村·七星峰

2-059 踏访梅溪：行，望，居，游（之七）（2015.11.18）
绍兴新昌，大桥，观音山，笔架山（金山）。

2-059 Visiting Meixi. Walking, viewing, living and travelling (7) (18 November 2015)

Big bridge, Guanyin Mountain, Bijia Mountain (Jin Mountain), Xinchang County, Shaoxing City.

2015.11.18. 大桥. 观音山. 笔架山.
（金山）

2-060 踏访梅溪：行，望，居，游（之八）（2015.11.18）

绍兴新昌，大水塘，七星山。

2-060 Visiting Meixi. Walking, viewing, living and travelling (8) (18 November 2015)

Qixing Peak, Big Pond, Xinchang County, Shaoxing City.

2015.11.18. 大水塘·七星山

2-061 踏访梅溪：行，望，居，游（之九）（2015.11.18）

绍兴新昌，胡卜村口（西南）。

2-061 Visiting Meixi. Walking, viewing, living and travelling (9) (18 November 2015)

Entrance to Hubu Village (Southwest), Xinchang County, Shaoxing City.

胡卜村口（西南）

2-062 踏访梅溪：行，望，居，游（之十）（2015.11.18）

绍兴新昌，胡卜村口（东）。

2-062 Visiting Meixi. Walking, viewing, living and travelling (10) (18 November 2015)

Entrance to Hubu Village (East), Xinchang County, Shaoxing City.

胡卜村口（东）

2-063 踏访梅溪：行，望，居，游（之十一）（2015.11.18）

绍兴新昌，胡卜村大会堂遗骸（址）。

2-063 Visiting Meixi. Walking, viewing, living and travelling (11) (18 November 2015)

Great Hall Ruin of Hubu Village, Xinchang County, Shaoxing City.

2015.11.18. 胡卜大会堂遗骸

2-064 踏访梅溪：行，望，居，游（之十二）（2015.11.18）
绍兴新昌，茶园。

2-064 Visiting Meixi. Walking, viewing, living and travelling (12) (18 November 2015)

Tea garden, Xinchang County, Shaoxing City.

2015.11.18 茶园

2-065 安徽绩溪，徽杭古道（2016.04.23）

故道田庄，烟雨沧桑。石尖欲落，泉喷谷响。

2-065 The past road linking Huizhou and Hangzhou in Jixi County, Anhui Province (23 April 2016)

The past road, field and village and misty rain tell the vicissitudes of life. The tip of a stone seems to be falling, and the sound of the spraying spring echoes in the valley.

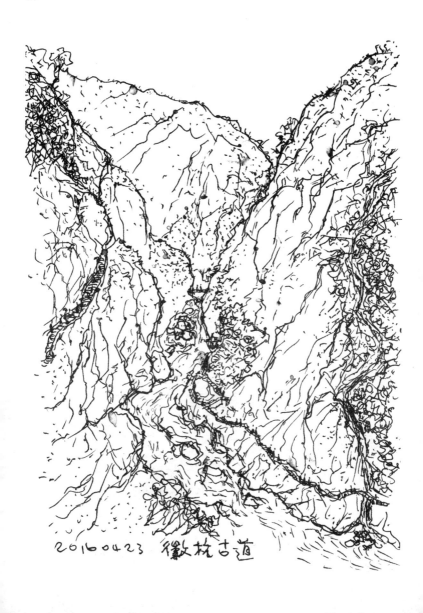

20160423 徽杭古道

2-066 安徽绩溪，尚村，水田（2016.04.23）

2-066 Water Field, Shang Village, Jixi County, Anhui Province (23 April 2016)

2-067 安徽绩溪，尚村，山野村庄（2016.04.23）

2-067 Shang Village fields, Jixi County, Anhui Province (23 April 2016)

2016.04.23 绩溪尚村·山野·村庄

2-068 浙江兰溪，诸葛村（2015.12.21）

2-068 Zhuge Village, Lanxi City, Zhejiang Province (21 December 2015)

2015.12.21. 南普村

2-069 黔西南，万峰林（之一）（2016.11.08）

抽象如盆景。

2-069 Wanfenglin Mountains, Southwest Guizhou Province (1) (8 November 2016).

Abstract as a bonsai.

2016.10.08 石峰林

2-070 黔西南，万峰林（之二）（2016.11.08）

万峰之林印象与地质景观原理。

2-070 Wanfenglin Mountains, Southwest Guizhou Province (2) (8 November 2016)

The imagery of "forest of ten thousands peaks" and the principle of the geological landscape.

2015. 11. 08　万山峰林 印象

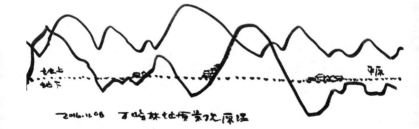

地上
地下

中原

2016.11.08 万暗林地质实况原理

2-071 黔西南，雨补鲁，清水河大峡谷（2016.11.08）
山水印象，意象。

2-071 Qingshuihe Grand Canyon, Yubulu Village, Southwest
Guizhou Province (8 November 2016)
　　Impression and imagery of the landscape.

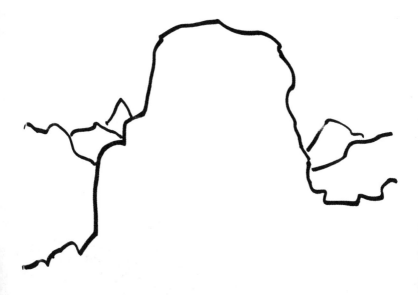

2019.11.08 清水河大峡谷 印象

2-072 洛阳龙门石窟（之一）（2016.01.02）

伊河东岸远眺西山（龙门山）。无数尊佛端坐凝视的山房。无论宏大与渺小，精微与粗陋，都是一个个完整的世界。过去、现在、未来并置于空间，隔岸而望，那竟是宇宙和自然的一个巨型剖面图。

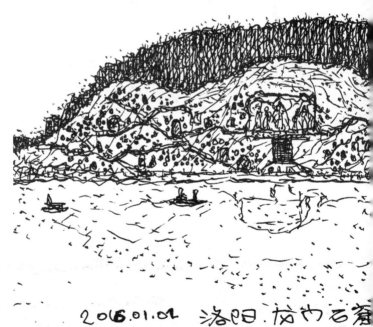

2-072 Longmen Grottoes, Luoyang (1) (2 January 2016)

Overlooking the Longmen Mountain in the west from the east bank of the Yihe River. Countless Buddhas stare from their seats in their mountain houses. Whether big or small, subtle or rough, they are all complete worlds. The past, present and future are juxtaposed in space. Looking across the shore, the grottoes become a giant cross-section of the universe and nature.

西山石 伊河东崖这此
龙门(山)

2-073 洛阳龙门石窟（之二）（2016.01.02）

自东山（香山）山谷望西山（天门山）。

2-073 Longmen Grottoes, Luoyang (2) (2 January 2016)

Looking at the Longmen Mountainin the west from the Xiang Mountain valley in the east.

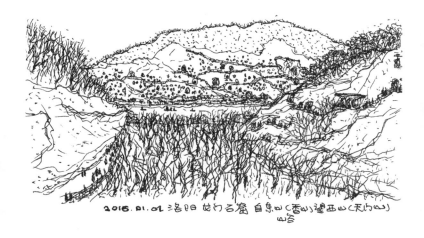

2016.01.02 洛阳 龙门石窟 自东山（香山）望西山（天门山）
西台

2-074 洛阳龙门石窟（之三）（2016.01.02）
奉先寺（大卢舍那像龛）。

2-074 Longmen Grottoes, Luoyang (3) (2 January 2016)
Fengxian Temple (Locana Buddha Statue Niche).

2016.01.02. 奉先寺（大卢舍那像龛），洛阳龙门石窟

2-075 五台山塔院寺 （2014.11.10）

寺前石桥水池，寺中白塔树丛，寺后庙群远山。

2-075 Tayuan Temple, Wutai Mountain (10 November 2014)

The stone bridge and pool in front of the temple, the white tower and trees in the temple, the temple group and the distant mountains behind the temple.

2014.11

2-076 太行山（之一）（2014.11.12）
自五台山往佛光寺途中，田畴，枯树，群山。

2-076 Taihang Mountain (1) (12 November 2014)
   On the way from Wutai Mountain to Foguang Temple: fields, dead trees and mountains.

2-077 太行山（之二）（2014.11.12）
自五台山返京途中，苍莽的山谷。

2-077 Taihang Mountain (2) (12 November 2014)
On the way back from Wutai Mountain to Beijing, a vast valley.

2014.11.12. 太行山（五台山返京途中）

2-078 太行山（之三）（2014.11.12）
　　山中村庄。

2-078 Taihang Mountain (3) (12 November 2014)
　　Village in the mountains.

2014.11.12. 太行山 森盼左

2-079 五台山佛光寺（2014.11.12）
自东大殿后山西眺寺群、原野、远山。

2-079 Foguang Temple, Wutai Mountain (12 November 2014)
Looking west at the temple group, the wilderness, the distant mountains from the hill behind East Hall.

2014.11.12. 佛光寺东大殿后山西眺

2-080 台湾花莲，太鲁阁燕子口隧道（之一）（2016.01.27）

"壶穴"内观：洞内有洞，道外有道，鬼斧神工。

2-080 Yanzikou Tunnel, Tailuge Mountain, Hualien County,
Taiwan (1) (27 January 2016)

    Inside view from the "pot hole": a hole inside a hole, a road
beyond a road, uncanny and artificial.

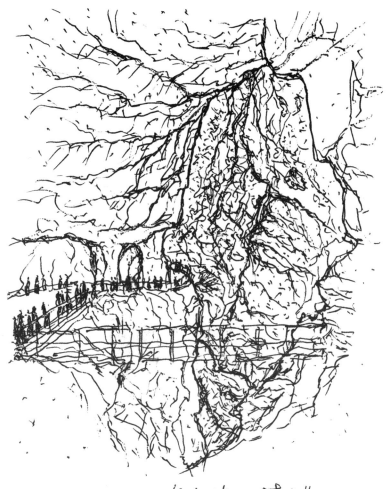

2016.01.27. 太舟阁燕子口"壶穴"
隧道

2-081 台湾花莲，太鲁阁燕子口隧道（之二）（2016.01.27）
"壶穴"外观：断崖通道，高岭幽谷，急流飞瀑，险峻大美。

2-081 Yanzikou Tunnel, Tailuge Mountain, Hualien City, Taiwan (2)
(27 January 2016)

Outside view of the "pot hole": passage on the cliff, high ridge and sequestered valley, rapids and waterfalls, steep and beautiful.

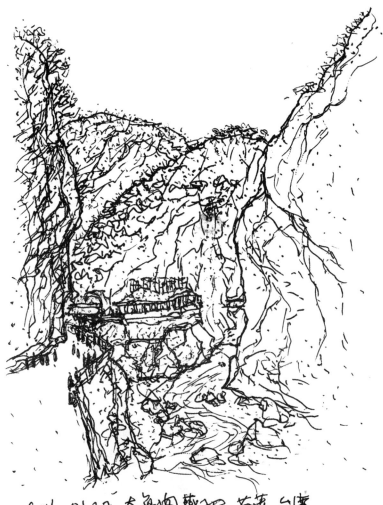

2016.01.27 太魯閣燕子口, 花蓮, 台灣
隧道藝術橋

2-082 北京，黄花城长城（之一）（2019.05.04）
夕阳下的长城，山谷，溪流，人家。

2-082 Huanghuacheng Great Wall, Beijing (1) (4 May 2019)
The Great Wall at sunset: valleys, streams, a household.

2019.05.04. 黄花城长城
夕阳下

2-083 北京，黄花城长城（之二）（2019.05.04）
烽火台上，透过残墙望远山。

2-083 Huanghuacheng Great Wall, Beijing (2) (4 May 2019)
On the beacon tower, looking out over the mountains through the residual walls.

2019.05.04. 黄花城长城

2-084 北京，黄花城长城（之三）（2019.05.04）
城墙随群山起伏。

2-084 Huanghuacheng Great Wall, Beijing (3) (4 May 2019)
The wall rises and falls with the mountains.

2019.05.04. 北京 慕田峪城 箭口长城

2-085 烟台，烟台山海岸 （2016.07.09）
近礁，船坞，远山。

2-085 Yantaishan coast, Yantai City (9 July 2016)
Near the reef, the dock, the distant mountains.

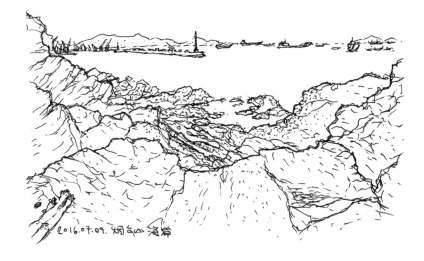

2016.07.09. 胡岛山海岸

2-086 威海，成山头好运角（2016.09.09）

夕阳之中，平远风景。

2-086 Cape of Good Fortune (*Haoyunjiao*), Chengshantou, Weihai City（9 September 2016）

The setting sun, far distance views.

R2016.09.09. 成山头 好运角

2-087 延庆，海坨山，佛峪口水库大坝 （2017.12.08）

萧瑟寒山。水坝像个美男子，安静优雅，兀自矗立。

2-087 Foyukou Reservoir Dam of Haituo Mountain, Yanqing (8 December 2017)

Bleak, cold mountains. The dam is like a handsome man, quiet and elegant, standing proudly.

2017.12.08 饱峪口水库，水坝俯拍。萧翠春山

2-088 飞机上俯瞰，墨西哥第一印象。此前就是电影《巴别塔》中有关墨西哥的场景（2017.01.09）

这个地方真干啊，巴拉巴拉的干，难怪他那么喜欢水的主题，有了水，就成了巴拉"甘"。

2-088 Overlooking Mexico from the plane, my first impression. My previous view of Mexico was in the movie *Babel* (9 January 2017)

This place is really dry; no wonder Luis Barragan likes the theme of water so much. With water, it becomes Barragan.

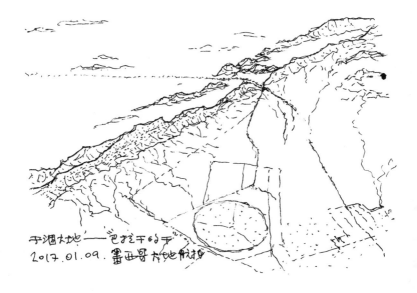

干涸大地——巴拉圭的干"
2017.01.09. 墨西哥大地航拍

2-089 飞机上俯瞰，墨西哥城市 （2017.01.09）

城市席卷自然。

2-089 Overlooking a Mexican City from the plane (9 January 2017)

The city follows a rolling topography.

2017.01.09. 墨西哥大地俯瞰 —— 土城市带着自然

2-090 德国国王湖雪景 （2018.01.22）

山雪漫漫。童话世界是真的存在的。

2-090 Snow view of King's Lake, Germany (22 January 2018)

With the deep snow in the mountains, the world of fairy tales is really there.

2018.01.22. 囫团湖 / 慕尼黑 / 德囻
山雪漫漫，童话世界

2-091 德国沃尔夫冈 （2017.10.19）

峰刃壁立：山相水相各不同。

2-091 Wolfgang Mountain (Lake), Germany (19 October 2017)

The peaks stand like blades: different appearances of mountains and water.

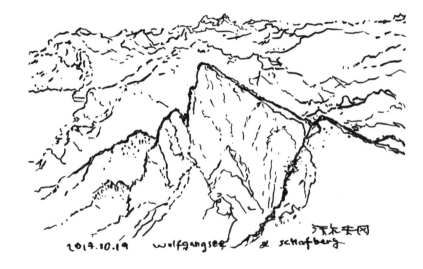

2017.10.19   wolfgangsee   & schafberg

2-092 英国科茨沃尔德（2013.07.16）

奇平坎普登小镇的小市场建筑，双廊，石柱＋屋顶木结构。

2-092 Cotswolds, UK (16 July 2013)

Small market building in the town of Chipping Campden:
double gallery, stone pillar and a timber structure roof.

CHIPPING CAMPDEN
Market Hall

2-093 英国温德密尔（2013.07.17）

山顶小石屋，生动、自然、安静的素人建筑。

2-093 Windermere, UK (17 July 2013)

A small stone house on the top of the mountain, a lively, natural and quiet building built by an amateur.

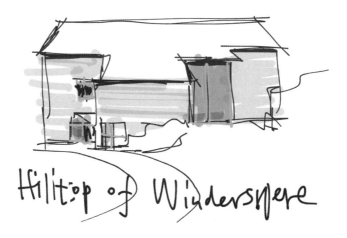

Hilltop of Windermere

2-094 英国苏格兰首府爱丁堡（之一）（2013.07.19）
自然与城市。

2-094 Edinburgh, Scotland, UK (1) (19 July 2013)
Nature and the city.

317

2-095 英国苏格兰首府爱丁堡（之二）（2013.07.19）
高地，都市胜景。

2-095 Edinburgh, Scotland, UK (2) (19 July 2013)
Highlands, urban poetic scenery.

2-096 英国苏格兰首府爱丁堡（之三）（2013.07.19）
深远都市。

2-096 Edinburgh, Scotland, UK (3) (19 July 2013)
Far-reaching city.

2-097 (1-2) 奥地利哈尔施塔特镇（2013.10.10）

山湖之间是城镇，栖居。

2-097 (1-2) Hallstatt, Austria (10 October 2013)

Between the mountains and the lake is a town. Dwellings.

2-098 捷克克鲁姆洛夫小镇（2013.10.11）

山地之城，地势高低与建筑组合，不断变化的街景。

2-098 Cesky Krumlov, Czech Republic (11 October 2013)

The mountain city, a combination of high and low topography with changing streetscapes.

ČESKÝ KRUMLOV   2013.10.11.

栖居于自然，无疑是一种人工对于自然的深度介入，使人获得身体与心灵、物质与精神层面的双重愉悦，实质是一种人工与自然互成的状态与情境，可称得上是人的生活空间营造的理想境界。人在自然中的栖居，形成了各种类型的聚落，并延伸为乡村与城市、建筑和园林等。多年以来的思考和研究逐步涵盖城市、建筑、园林、聚落之后，我越来越认识到一种存在于四者之中的"一体性"：它们都不过是中国传统营造体系的不同方面与不同方式的存在和呈现，基于相同的生活哲学，亦即对建筑与自然之间更为紧密互动、相互依存、共生共长之关系的格外关注，一种与特定人群生活理想密切关联之状态的营造，它们均可被视为"广义的聚落"。

我对中国传统营造体系的最初感知始于城市，始于在建筑启蒙时期对北京故宫（紫禁城）的"平远"观望与巨大心灵触动，继而延伸至建筑、园林、聚落。关于理想世界、理想居所、理想城市、理想建筑（"晋东南五点"和"蔚县五点"等）、理想聚落乃至"现实理想空间营造范式"（"佛光寺五点范式性启示"）的感悟心得，几乎都是来自旅途中对城市、建筑、园林、聚落的实地体望、触发、思考和总结。

在这样观想和体悟的思考历程中，来自中国传统之外的他者及当代案例亦构成重要的平行参照和启示。

Viewing
(Settlements, cities and buildings)

Living in nature is undoubtedly a deeply artificial intrusion. It enables people to obtain the dual pleasures of the body and mind, gaining both material and spiritual benefits. The essence is the inter-state or the situation between the artificial and the natural, which can be called the ideal realm for building human living space. Human habitation in nature is formed of various types of settlements that extend into villages and cities, buildings and gardens. After years of thinking about and researching cities, buildings, gardens and settlements, I have gradually become more aware of the integration that exists between the four: based on the same philosophy, life and thoughts, they are the existence and presentation of different aspects and manifestations of the traditional Chinese construction system. They jointly demonstrate a special focus on the close, interdependent and symbiotic relationship between architecture and nature, and they are closely related to the life ideals of a particular group of people. Therefore, they can all be regarded as generalised settlements.

My initial perception of the traditional Chinese construction system originated with cities, then extended to architecture, gardens and settlements. Specifically, Beijing's Forbidden City, with its spectacular roofscape and far-distance views, gave me great spiritual energy during my architectural enlightenment. My feelings about and experience of the ideal world, homes, cities, architecture (the five points of southeastern Shanxi Province and the five points of Yu County), settlements and even the paradigm of building realistic ideal space (Foguang Temple's five-point paradigm revelation) almost came from thinking about and summarising my engagement with cities, buildings, gardens and settlements during many trips.

In such a process of thinking, other people and contemporary cases from outside the Chinese tradition also constituted important parallel references and revelations.

3-001 北京，自景山万春亭俯瞰紫禁城和北京城(2017.01.29)

"胆敢独造"：工具、材料、技法、语言、形神、意境、情感的探寻——观北京画院"清寂鹜影"之林风眠展＋"何要浮名"之齐白石展。返程路过景山，在寒风及人喧中速写一页紫禁城。

距离 1990 年夏天第一次爬上景山见到这个场景，时间已过去近 30 年，内心的感触与叹息依然新鲜如昨。当年由此起点出发，我仿佛已历经千山万水，"归来仍如少年"，而所有的一切沧桑、一切思考尽在其中，尽在眼前。北京和故宫是我的起点，也是我的不断回归之点。

3-001 Overlooking the Forbidden City and Beijing City from Wanchun Pavilion, Jingshan Hill, Beijing (29 January 2017)

Dare to create (*dangan duzao*): tools, materials, techniques, languages, shapes, spiritual/artistic conceptions, the exploration of emotions. After visiting the Lin Fengmian Exhibition, *Lonely Shadow* (*jingji wuying*), and the Qi Baishi Exhibition, *Why to Be Famous* (*heyao fuming*), shown at the Beijing Academy of Painting, when passing Jingshan Hill on my return journey, I drew a sketch of the Forbidden City in the cold wind and crowds.

It has been nearly 30 years since I saw this scene for the first time in summer 1990. The inner feelings and awe are still as fresh as if it happened yesterday. Starting from this, I seem to have travelled for thousands of miles and returned as if still a teenager. All of the vicissitudes of life, all of my thoughts are in front of me. Beijing and the Forbidden City are my starting point and a point of constant return.

3-002 晋东南，泽州岱庙（2014.06.29）

<u>风水形势</u>：建筑的选址、空间布局与外部自然环境的密切关联。

2014年6月，随天津大学、香港大学师生考察晋东南古建筑，于现场有感而发，小结了中国悠久深厚的营造传统的几个重要特征，称为"晋东南五点"：一、风水形势；二、庭院深奥；三、空间结构；四、屋顶之美；五、预制建造。试图归纳中国传统营造方式在环境、空间、结构、形式、建造等方面呈现出来的"物质性"特征。

3-002 Dai Temple, Zezhou, southeastern Shanxi Province (Jin) (29 June 2014)

<u>Feng shui situation</u>: the location and spatial layout of the buildings are closely related to the external natural environment.

In June 2014, I visited the ancient buildings in southeastern Shanxi Province with teachers and students from Tianjin University and the University of Hong Kong. I had some feelings on site and summarised several important characteristics of China's long and profound tradition of design and construction known as the five points of the southeastern Shanxi Province: (1) Feng shui situation; (2) deep courtyard; (3) spatial structure; (4) the beauty of the roof; (5) prefabricated construction. I try to sum up the physical characteristics of traditional Chinese design and construction methods in terms of the environment, space, structure, form and construction.

2019.06.29 于扬州瘦西湖

3-003 晋东南，长治县玉皇观（2014.06.30）

庭院深奥：由群体建筑围合的庭院，层层展现出动人的空间和景观。

3-003 Yuhuang Guan Temple, Changzhi County, southeastern Shanxi Province (Jin)(30 June 2014)

Deep courtyard: a courtyard enclosed by a group of buildings, showing layers of appealing spaces and landscapes.

3-004 晋东南，高平游仙寺大殿（2014.06.29）

　　<u>空间结构</u>：建筑的内部真实、合理的结构表达与空间营造。

3-004 Hall of Youxian Temple, Gaoping City, southeastern Shanxi Province (Jin)(29 June 2014)

　　<u>Spatial structure</u>: the real and logical structural representation and creation of space within the interior of a building.

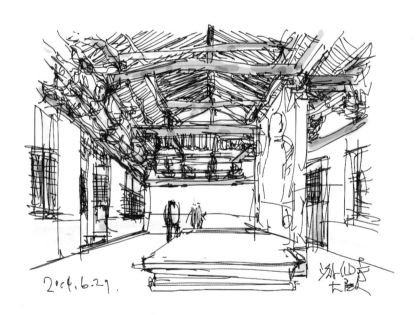

2004.6.29.

3-005 晋东南，长子县天王寺（2014.06.30）

屋顶之美：外部的建筑形式，比例、尺度、细部，特别是屋顶的视觉呈现。

3-005 Tianwang Temple, Changzi County, southeastern Shanxi Province (Jin)(30 June 2014)

<u>The beauty of the roof</u>: the external architectural form, proportions, scales and details, especially the visual representation of the roof.

2016. 6.10 善化寺大雄宝殿

3-006 晋东南，长子县汤王寺大殿木构架（2014.06.30）

<u>预制建造</u>：一种预制建造体系，对于设计、建造、使用和维护的重要意义和启发性。

3-006 The wooden frame of the Tangwang Temple in Changzi County, southeastern Shanxi Province (Jin)(30 June 2014)

<u>Prefabricated construction</u>: a kind of prefabricated construction system, important and inspiring for design, construction, use and maintenance.

3-007 蔚县，南安寺塔（辽）（2015.08.06）

<u>气质与精神</u>（辽的气质）：一种外部与内在的高妙组配。通过建筑形式和内部结构、空间相互匹配的整体精神来营造出建筑的独特气质。

2015 年 8 月和 2016 年 8 月两次随天津大学师生到蔚县考察和测绘，又获得"蔚县五点"：一、气质与精神；二、全息的诗题；三、废墟园庭；四、现象与现实；五、不断展开的山水。这五点分别聚焦于中国传统建筑的文化性、文学性、自然性、人类性以及叙事性等精神性特征。

3-007 Nan'an Temple Tower (Liao Dynasty), Yu County (6 August 2015)

<u>Temperament and spirit</u> (the temperament belongs to the Liao Dynasty): this tower is a perfect combination of exterior and interior. The unique temperament of the building is created by the overall integration of its architectural form and its internal structure and space.

In August 2015 and August 2016, I went to Yu County with faculty and students from Tianjin University to study and survey historic buildings, and then summarised the five points of the local architecture: (1) temperament and spirit; (2) the "holographic poem"; (3) the ruined garden; (4) phenomenon and reality; (5) continously unfolding landscape. These points focused respectively on the spiritual characteristics of Chinese traditional architecture, namely culture, literature, humanity, nature and narrative.

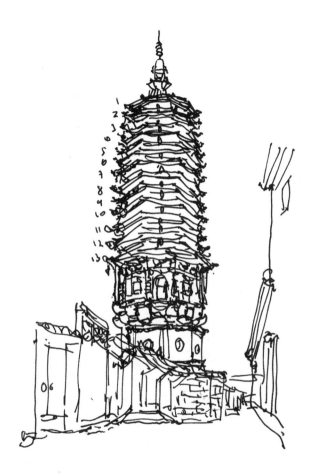

341

2015.08.06.
尉氏 南安寺塔 (元) 川年
1-17066建

3-008 涞源，阁院寺文殊殿 （辽） （2015.08.13）

双树，门窗（梵字），结构，斗拱，观看距离（进深×2＝殿前距离）。

3-008 Wenshu Hall (Liao Dynasty) of Geyuan Temple, Laiyuan
County (13 August 2015)

Double trees, doors and windows (Sanskrit), structure, *dougong*
and viewing distance (building depth × 2 = distance of the front
temple).

3-009 涞源，阁院寺文殊殿 背（北）立面檐下速记（2015.08.13）
被笼罩感。

3-009 Sketch of Wenshu Hall of Geyuan Temple under the eaves of
the back (north) facade, Laiyuan County (13 August 2015)
　　Being enveloped.

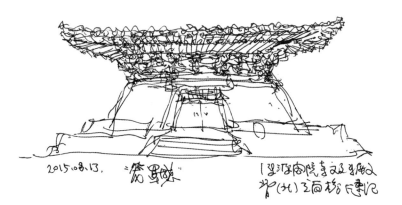

3-010 新城，开善寺大雄宝殿（2015.08.06）
内部空间结构。

3-010 The Great Buddha's Hall of Kaishan Temple, Xincheng
County (6 August 2015)
Internal spatial structure.

武城县开善寺大雄宝殿
2015. 08. 06.

3-011 蔚县，玉皇阁 （2015.08.08）

<u>全息的诗题</u>：一种情境的呈现。"目穷千里，云蒸霞蔚，历古阅今。"将人文性赋予自然中的建造物之上，成为"洞穿时空的利器"。

3-011 Yuhuang Pavilion, Yu County (8 August 2015)

<u>Holographic poem</u>: the presentation of a situation. Seeing for a thousand miles, the slowly rising rosy clouds, knowing the ancient and the present (*Mu qiong qian li, yun zheng xia wei, li gu yue jin*). The humanity that is attached to the buildings becomes a weapon to penetrate time and space.

自有千里
云蒸霞蔚
历古间今

2015.08.08.

碧玉流图

3-012 蔚县，任家庄，黍田与堡墙、远山 （2016.08.11）

"彼黍离离，心忧何求。"丁垚站在龙王庙残存的遗迹上，手握一把黍，让同学面向人群高声吟诵《诗经·黍离》，场景真切而感人，使人仿佛穿越到《诗经》时代的现场，古今通感。

3-012 Renjiazhuang Village, field and fort wall, distant mountains, Yu County (11 August 2016)

Ding Yao stood in the remains of the Dragon King Temple (*Longwangmiao*), held a handful of millets and let the students recite the *Book of Songs — Shuli* loudly. The scene was really touching, seemingly sending people back to the site of the *Book of Songs* in ancient period. Ancient and modern feelings are interlinked.

2016.08.11 卧马作永在 泰田与堡墙这山

3-013 蔚县，东庄头村，龙王庙戏台（2016.08.13）

古今悲欢。永远。复得返自然。"把古今事重新提起，将悲观情再现出来。"（檐下对联）

3-013 Longwang Temple Stage, Dongzhuangtou Village, Yu County (13 August 2016)

Ancient and modern, sorrow and joy, forever. Return to nature. Re-acknowledging ancient and modern affairs and representing pessimism (*Ba gujin shi chongxin tiqi, jiang beiguan qing zaixian chulai.*) (couplet under the eaves).

2016.08.13 蒙自县东在沙村
大风庙站台

把古今事重新提起
将悲欢情再说出来

3-014 蔚县水涧子，水东堡 （2015.08.07）

废墟园庭（废墟与败壁）："废墟" 是一种人工在自然中的消极退化，而"废墟园庭"则呈现出一种人工与自然的积极互动、组构与升华。

3-014 Shuidong Fortress, Shuijianzi Village, Yu County (7 August 2015)

The ruined courtyard-garden (ruins and dilapidated walls): ruins are a kind of artificial degradation by nature, whereas a ruined courtyard-garden presents a kind of active interaction, and sublimation of the artificial and the natural.

2015-08-07 废墟与败壁　　蔚县水山田了水马堡

3-015 蔚县水涧子，水东堡（2015.08.06）
长满荒草的庭院。

3-015 Shuidong Fortress, Shuijianzi Village, Yu County (6 August 2015)
Grassy courtyard.

355

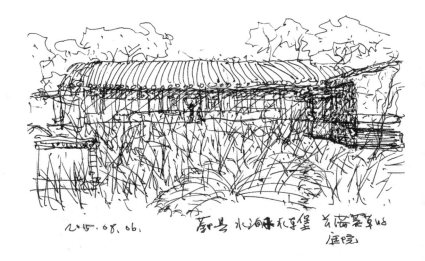

2015.08.06.　黔县 水涧味花事堡 芳满琴事的
庄院

3-016 蔚县，白家东堡 （2016.08.07）

"黄土庭园"。

3-016 Baijiadong Fortress, Yu County (7 August 2016)

Yellow earth courtyard.

2016.08.07 蔚县白家东堡 "黄土庭园"

3-017 蔚县，薄（卜）南堡（2015.08.09）
村庄前仅存的黄土败壁，犹如独石。

3-017 Bo(Bu)nan Fortress, Yu County (9 August 2015)

The only remaining loess wall in front of the village is like a monolith.

3-018 蔚县，任家庄军堡遗迹（之一）（2016.08.11）

平远地景。

3-018 Renjiazhuang Village Military Fortress Remains, Yu County,
(1) (11 August 2016)

Landscape with level distance.

2016.08.11. 蔚号庄家庄军堡

3-019 蔚县，任家庄军堡遗迹（之二）（2016.08.11）
山野深远。

3-019 Renjiazhuang Village Military Fortress Remains, Yu County,
(2) (11 August 2016)

The mountains are deep and distant.

3-020 蔚县盆地东部田野印象（2016.08.11）

田地中的下沉式"微缩"丘陵景观（犹如黄土山地盆景）。

3-020 Field impression of the eastern part of the Yu County Basin (11 August 2016)

The sunken "miniature" mountain and forest in the field (like the bonsai of the loess mountain).

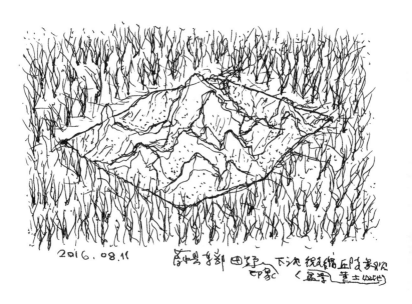

2016.08.11　　　廬山雷部 田野，下流 織絲繡丘陵景观
印象C　　（盛夏 葦土山地）

3-021 蔚县水涧子，中堡（2015.08.07）

现象与现实：一种更具包容性的建筑学。对建筑的观察、思考和认知，在通常局限于形态、空间、结构、材料、建筑的物质性和精神性等问题的建筑学"本体"之外，容纳更多的内容，如地理、文化、历史乃至"人类"。所有的聚焦点背后其实都是人，也就是人在特定的时间和空间中的存在及其对人工营造的影响。

3-021 Shuijianzi Village, Zhong Fortress, Yu County (7 August 2015)

Phenomenon and reality: a more inclusive architecture. In addition to the normal architectural ontology-morphology limited in space, structure, materials, physicality and spirituality, the general content such as geography, culture, history and even humanity is also contained. Behind all the focus points are people; that is, the existence of people in a specific time and space and their impact on artificial construction.

蔚县. 水涧子. 中堡 2015.08.07.

3-022 蔚县水涧子，水东堡 （2015.08.07）

由戏台外望，文昌阁，堡墙，堡门，观音庙，一组具有当代意味和品质的公共空间。

3-022 Shuijianzi Village, Shuidong Fortress, Yu County (7 August 2015)

Looking outside from the stage: Wenchang Pavilion, fort wall, fort gate and Guanyin Temple. This is a group of public spaces with contemporary meaning and quality.

2015.08.07

黄岩 水心街及水乡建筑地, 文昌阁
住宿, 建议, 双吾楼

3-023 蔚县暖泉镇，自老君观俯瞰 （2015.08.09）
以"人类学视角"登高眺远。

3-023 Overlooking from Laojunguan Temple, Warm Spring Town (*Nuanquanzhen*) , Yu County (9 August 2015)
Standing high and looking far away from an anthropological perspective.

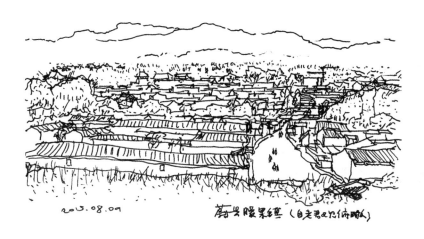

2015.08.09

蔚县暖泉镇（自走君又た俯瞰）

3-024 蔚县重泰寺，山体前序（2015.08.09）

不断展开的山水：行望居游，一种理想世界的组构与体验。

进入寺庙之前那些大大小小、高高低低的土山与人工步道、台阶、丘壑相互掩映的关系，就像是寺庙空间体验的前序，如果将其视为一个刻意的营造，那将是非常高妙的设计，算是"小山小水"。

3-024 Chongtai Temple, front sequence to the mountain (9 August 2015)

<u>Continuously unfolding landscape</u>: walking, viewing, living and wandering, the composition and experience of an ideal world.

The large and small, high and low dirt hills and the artificial trails, steps and hills creates a unique spatial experience and sequence approaching the temple. If this could be considered a deliberate creation, it would be a very wonderful design, or "small-scale mountain landscape".

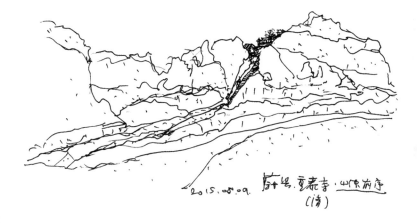

2015.08.09. 莒县 浮来寺·山东省莒县
(清)

3-025 蔚县重泰寺，中途丘壑景象（2015.08.09）

3-025 Gully landscape seen on the way, Chongtai Temple, Yu County (9 August 2015)

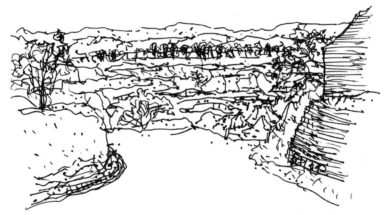

2015.08.09. 崇善 重乳寺(清)

3-026 蔚县重泰寺，庭园中的树与房 （2015.08.09）

3-026 Tree and house in the courtyard, Chongtai Temple, Yu County (9 August 2015)

2015.08.09

3-027 蔚县小五台金河寺庭院（2015.08.10）

　　一处平地上的建造，重要的是前面整个的进入体验过程——经历空间的开合、高低和明暗的变化，最后到达寺庙，可称作"游山玩水"。

3-027 Courtyard of Jinhe Temple, Xiaowutai Mountain, Yu County (10 August 2015)

　　For this building built on flat land, the important thing is the whole process of entering: experiencing the opening and closing of space, the change of height and brightness and finally reaching the temple. This can be called "sightseeing among hills and rivers" (*youshan wanshui*).

2015.08.10. 나로남 통도사
(畫)

3-028 俯瞰蔚县小五台金河寺及塔群 （2015.08.10）

3-028 Overlooking Jinhe Temple and the group towers in the
Xiaowutai Mountain, Yu County (10 August 2015)

2015.08.10

青海 金洞寺(?) 風塔路

3-029 太行小五台印象 (2015.08.10)

经历一路艰难险阻的游览过程，体会所有那些在山水画中的高远、深远、平远空间以及景象的变化，实际上两者都是一个不断被串联起来的、对空间和景象的体验过程。回来后闭门造车，给小五台的整个游览过程的意象画了一幅"立轴"草图，实际上并非真实的相机拍照得到的或者眼睛看到的完整画面，它更是一种人工的构造和组合——就像建筑师做的设计，把所有的场景组合在一个空间中——串联起所有的观察，亦即中国特有的"散点透视"山水画作。这是第一次用这样特殊的方式绘画山水草图，我称之为"不断展开的山水"。

3-029 Impression of Xiaowutai Mountain, Taihang Mountain (10 August 2015)

On this journey, full of hardships and obstacles, I experienced the spatial changes and changes in scenery in landscape paintings, such as the ways of portraying high distance, deep distance and level distance. It was a continuous interweaving process of experiencing spaces and scenes. After I returned, I shut myself up in a room and drew a sketch with a "vertical axis" of the entire tour of Xiaowutai Mountain. It was not a complete or real picture taken by a camera or seen with the eyes, but an artificial structure and composition, just like the architect's design, composing all the scenes and observations in one space, similar to China's unique scattering perspective of landscape paintings. This was the first time I used this special way to draw a landscape sketch, which I call the "continuously unfolding landscape".

太行小五台印象

2015. 08. 10

3-030 蔚县地区地形及村堡分布图 （2016.08.07）

理想聚落：往日的堡村，今日的遗迹，明日的庙宇。聚落，随着时间而兴而衰，在自然与人工之间相互转化。但在大的地理图像中，所有的它们都不过是盖在自然几何之上的一个个大大小小、各形各状，或简单或复杂、或清晰或模糊的戳儿。

3-030 Map of the topography and village forts in Yu County (7 August 2016)

Ideal settlement: the village fort of the past, the remains of today, the temple of tomorrow. Human settlements rising and falling with time have experienced the transformation of both nature and artefacts. However, viewed within large geographic images, all of them are just a few large or small, simple or complex, clear or fuzzy stamps with various shapes covering the natural landscape.

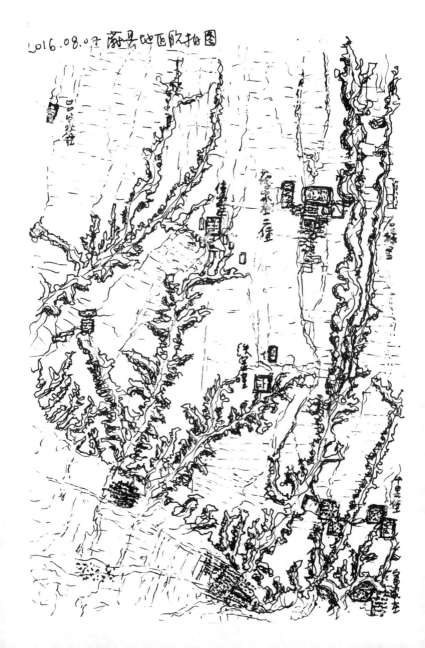

3-031 五台山，龙泉寺平面图（2014.11.11）

小（s）、中（m）、大（l）、超大（xl）四个庭院建筑一字排开，自如缩放的尺度令人吃惊；建筑布局与地形的匹配安排自由而有机；充分运用墙体、栈道、平台等景观（非建筑）元素。如同大师之作，相信应是"素人"营造。

3-031 Plan of Longquan Temple in Wutai Mountain (11 November 2014)

The four buildings, with small, medium, large and extra-large courtyards are lined up in a row. This freely changing scale of courtyards is surprising. The relationship between the architectural layout and the terrain is free and organic, with the full use of walls, plankings, platforms and other landscape (non-building) elements. I believe that this temple, like a masterpiece, was created by amateurs.

383

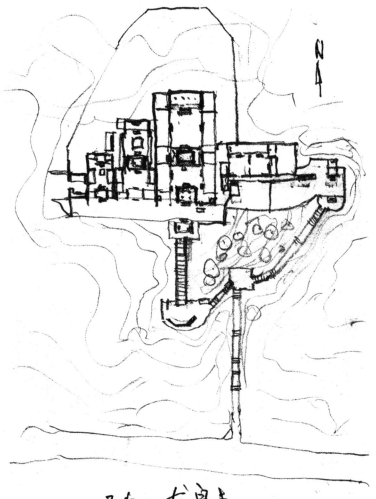

五台山龙泉寺.

2016.11.16

3-032 五台山，龙泉寺，寺后栈道与后院 （2014.11.11）

　　栈道沿着山坡百折千回，让开丛丛树林，连起位于各个高度的平台院落，格外自然的"素人"营造。

3-032 The plank road and backyard in Longquan Temple, Wutai Mountain (11 November 2014)

　　The plank road runs back and forth along the hillside, avoiding the clumps of woods and connecting platforms and courtyards at various heights. It is an exceptionally natural "amateur" creation.

2016.11.11 星台山的兔夺 李拍线直山间兔

3-033 平遥，南城门，兴国寺遗迹景观（2016.06.10）

作为世界文化遗产的平遥古城，原本的市民生活空间已基本搬迁到城外，剩下的只是大大小小的空间躯壳，和无所不在令人窒息的旅游及商业。只有靠近南城门内的兴国寺模拟复原遗迹，夕阳时分，城墙城楼之下，游人市民在此休闲，烤串香气诱人，颇有几分真正的生活烟火之气，令人动心动容。

3-033 Relics of Xingguo Temple, South Gate, Pingyao City (10 June 2016)

In the ancient city of Pingyao, a UNESCO world cultural heritage site, the original living space for citizens has been basically moved outside the city, leaving only large and small empty spaces full of omnipresent tourists and businesses. In Xingguo Temple restored relics near the south gate of the city, at sunset, under the city wall and tower, some tourists and citizens relax here. The enticing aroma of barbecued skewers showcases quite a lively and touching scenario.

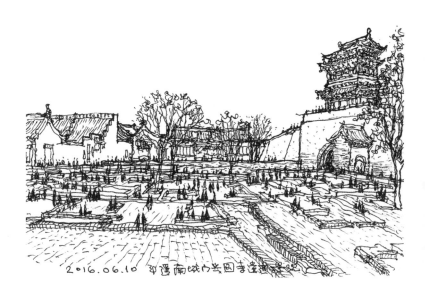

2016.06.10 邓峰南城内兴国寺造像碑

3-034 平遥，镇国寺（2016.06.10）

昨天光顾吃烤串，忘了附城图，今昔对比，不胜叹息，城市以前严密完整的社会、生活系统，如今已碎片化、景观化、单一商业化，忽然想若是当年"梁陈规划"果真落地，北京旧城会不会也被弄成这样。另平遥城内并无如北京三海的自由水系，也无西北山脉依傍屏障（所谓"北京湾"），北京之址建城真乃得天独厚，"山水城市"是也，又叹。城外镇国寺（五代），跟平遥古城一起入"世遗"，名副其实。

3-034 Zhenguo Temple, Pingyao City (10 June 2016)

Yesterday, I was eating barbecued food and forgot to attach the city map. The contrast between the past and the present is remarkable. The city's previously rigorous and complete social and living system has now become a fragmented and commercialised landscape. Suddenly, I begin to wonder whether the old city of Beijing would be like this if the Liang-Chen Planning Proposal had been realised. In Pingyao City, there is no free water system like the three seas in Beijing, and no barrier similar to the northwestern mountain range (the so-called Beijing Bay). The site of Beijing is truly unique, and in this sense, it is a city with a mountain and river (*shanshui chengshi*). However, Zhenguo Temple outside the town (built during the Five Dynasties) and the ancient city of Pingyao, listed as world cultural heritage sites, are worthy of their reputations.

2016.06.18 苏州西园寺

3-035 灵石，王家大院 （2016.06.11）

祁县乔家大院和灵石王家大院，大红灯笼高高挂，没完没了的院子。

3-035 The Wang's Grand Courtyard, Lingshi County (11 June 2016)

In the Qiao's Grand Courtyard of Qi County, and the Wang's Grand Courtyard of Lingshi County, the red lanterns are hung high in the endless courtyards.

2016.06.11 晋古王家大院

3-036 灵石，王家大院红门堡东望（2016.06.11）

在山地林木的包裹之中，窑洞地坑院和大院的结合应是大院的"前世"类型。

3-036 Looking east from the Hongmen Fortress to the Wang's Grand Courtyard, Lingshi County (11 June 2016)

The combination of underground cave dwellings and courtyards surrounded by mountains and forests should be the precedent type of compound (*dayuan*).

2016.06.11 王家大院 红□堡东望

3-037 摹写冯建逵先生《悬空寺》图（2019.06.26）

　　山西浑源城南之悬空寺，始建于北魏太和十五年（491 年）。崖寺相依，"上载危岩，下临深谷，半插飞梁为基，巧借岩石暗托，布局参差有致，飞栈暗道勾连，仅 152.5 平方米地基上，建有大小殿阁 40 余间，是国内唯一现存佛道儒三教合一的寺庙"（张兵，恒山悬空寺，《文史月刊》2016 年第 11 期，p51）。距地 80 余米悬空而立，经历百次地震、千年风雨，是北岳恒山十八景中"第一胜景"。清乾隆《浑源州志》记述："悬空寺在州南恒山下磁窑峡。悬崖三百余丈，崖峭立如削，（寺）倚空凿窍，结构层楼，危梯仄蹬，上倚遥空，飞阁相通，下临无地，恒山第一景也。"

3-037 Imitating Feng Jiankui's The Hanging Temple (*Xuankongsi*) (26 June 2019)

　　The Hanging Temple located in the south of Hunyuan City, Shanxi Province was built in 491. The cliff and the temple are interdependent. "The temple, along a deep valley below, is bearing high rocks on the top. It is based on beams partly inserted into the mountain. The builders used rocks smartly to secretly support it. The layout is uneven but orderly. The buildings are connected by outside and inside trestle roads along cliff. Built on only 152.5 square meters of foundations the temple has more than 40 large and small rooms and is the only existing temple with the combination of three Chinese traditional religions: Buddhism, Taoism, and Confucianism." (see Zhang Bing, "Hanging Temple in Hengshan Mountain", *Wenshi yuekan*, No. 11, 2016, p. 51). Suspended more than 80 meters above the ground, it has experienced hundreds of earthquakes, thousands of years of wind and rain, and is recognised as the first of the 18 scenic spots in the Hengshan Mountain in the north.

2019.06.26 蒙多乃建造先生 景区画图

3-038 正定，临济塔（2017.10.05）

正定隆兴寺，转轮藏殿，宋碑，四塔，心称其道也。

3-038 Linji Tower, Zhengding County (5 October 2017)

I admire the design and construction of the Longxing Temple, the Zhuanluncang Hall, the Song Dynasty stele and the four towers in Zhengding County.

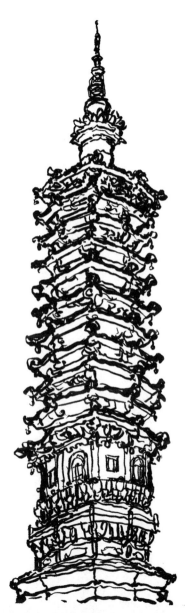

2017.10.05
正定临济塔

3-039 北京香山，香山寺、来青轩、鹫峰全图（2018.02.23）

3-039 The full map of Xiangshan Temple, Laiqingxuan Pavilion and Jiufeng Peak, Fragrant Hill (*Xiangshan*), Beijing (23 February 2018)

2018.02.23 北京香山寺. 栗宪廷

3-040 北京香山，香山寺后苑 （2018.02.22）

香山，香山寺，来青轩（遗迹），见心斋（未进入）。
2018"眼界宽"。

3-040 Backyard of Xiangshan Temple, *Xiangshan*, Beijing (22
February 2018)

Xiangshan, Xiangshan Temple, Laiqingxuan Pavilion (relics),
Jianxinzhai Building (haven't enterd). I hoped to have a wider
horizon in 2018.

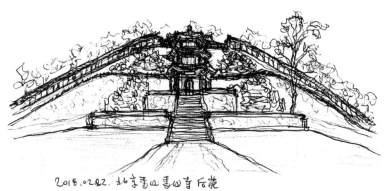

2018.02.22. 北京香山香山寺后苑

3-041 易县，清西陵全景示意图（2018.10.04）

风水形势。

3-041 Panorama of the Western Qing Tombs, Yi County (4 October 2018)

Feng shui situation.

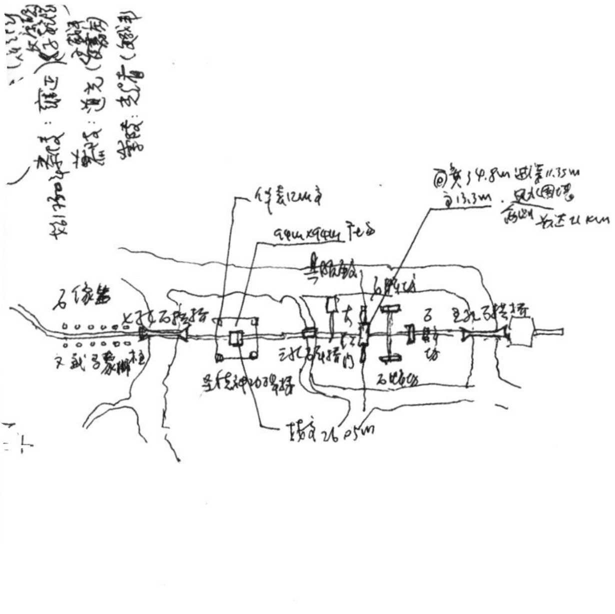
石像生

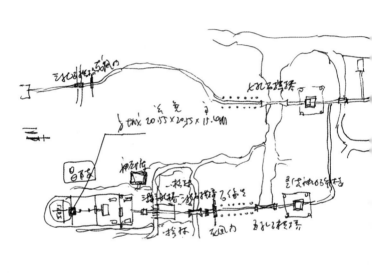

3-042, 043, 044

易县，清西陵，泰陵与昌陵平面示意图 （2018.10.04）

    "形"与"势"的对应尺寸与尺度。面宽、进深、高度，广场，方城，风水围墙。神道绕山而行。

3-042, 043, 044 Plans of Tai Tomb and Chang Tomb of the Western Qing Tombs, Yi County (4 October 2018)

    The size and scale of shape (*xing*) and the size and scale of potential (*shi*). Width, depth, height, square, quadrate city, feng shui bounding wall. The path leads to the tomb (*shendao*) and goes around the mountain.

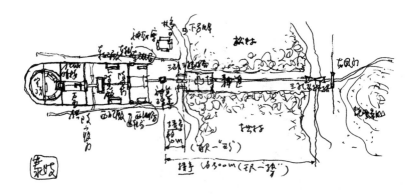

3-045 易县，清西陵，泰陵，龙凤门 （2018.10.04）

门对神道殿宇的过白。

3-045 Longfeng Gate of the Tai Tomb, Western Qing Tombs, Yi County (4 October 2018)

The gate creates a *guobai*, framing the path and the temple.

3-046 易县，清西陵，泰陵，龙凤桥（2018.10.04）
拱桥面和栏杆对殿宇的过白。

3-046 Longfeng Bridge, Tai Tomb of the Western Qing Tombs, Yi County (4 October 2018)

The arched bridge and the railings create a *guobai*, framing the temple.

3-047 易县，清西陵，泰陵，龙凤桥上（2018.10.04）

因"空间压缩"的作用，神道尽端的殿宇屋顶群看上去"贴"
在了一起，形成完美的构图。

3-047 On the Longfeng Bridge, Tai Tomb of the Western Qing
Tombs, Yi County (4 October 2018)

　　Due to "space compression", the group of temple roofs at
the near end of the path seem to be combined, forming a perfect
composition.

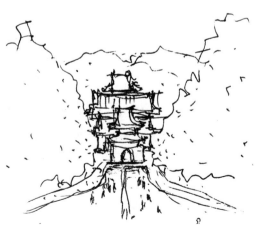

3-048 易县，清西陵，泰陵，凸形神道上望殿宇群(2018.10.04)

神道尽端的一段百米凸形甬道，使得人在行进中，前方殿宇屋顶群保持"贴"在一起的完美构图"不动"。是谓"步移景滞"。

3-048 Looking at the group of temples on the *shendao*, Tai Tomb of the Western Qing Tombs, Yi County (4 October 2018)

Walking on the 100-meter-long raised paved path, one can find a perfect "fixed" composition of the front roofs fused together. This is called "moving body but seeing the same scenery".

3-049 易县，清西陵，泰陵，凸形神道尽端望殿宇群

（2018.10.04）

　　直到甬道尽端，碑亭广场蓦然出现，殿宇群一下子"散开"为正常的空间构成。多么高妙的设计！

3-049 Looking at the group of temples from the proximal end of the *shendao*, Tai Tomb of the Western Qing Tombs, Yi County (4 October 2018)

　　Arriving at the end of the paved path, one immediately sees the square with the tablet and pavilion, and the group of temples is suddenly spread out, forming a normal spatial composition. What a wonderful design!

2018.10.04. 泰陵必朝神道上. 清西陵

3-050 易县，清西陵，泰陵（2015.08.13）

山川田野，人工水系，神道殿宇，风水营造。

3-050 Tai Tomb of the Western Qing Tombs, Yi County (13 August 2015)

Mountains and fields, the man-made water system, path and temple, *feng shui* design and construction.

2015.08.13.

男号1 庄陵 素陵

3-051 遵化，清东陵，定陵 （2019.06.13）

三十年后再来遵化清东陵。登定陵方城自北望南，天台金星朝山圣境。

3-051 Ding Tomb of the Eastern Qing Tombs, Zunhua City (13 June 2019)

After thirty years, I came to Eastern Qing Tombs in Zunhua again, climbing the quadrate gate tower of the Ding Tomb and looking from north to south. These tombs face Tiantai Mountain and Jinxing Mountain. Such a situation creates a holy place.

2019.06.13 诸暨勾践·兰陵. 朝山为天台山，左边为会稽山

3-052 遵化，清东陵，由金星山顶北侧"平台"俯瞰陵区

（2019.06.13）

　　中央神道轴线南北长 7 公里，长度相当于从北京城安定门至永定门，串联起广袤的田野、河道、村庄、人家，两翼群山环抱，生机勃勃。

3-052 Overlooking the mausoleum by the platform on the north side of Venus Peak, the Eastern Qing Tombs, Zunhua City (13 June 2019)

　　The 7 km north-south central axis of the *shendao*, equivalent to the distance from Anding Gate to Yongding Gate in Beijing, is connected to the vast fields, rivers, villages and houses. The two wings of the *shendao* are surrounded by mountains and are full of vitality.

2019.06.13. 落采日暮   由金界昆山俯瞰台陵区

3-053 遵化，清东陵，风水形势图（2019.06.13）

气喘吁吁第一次爬上金星山，自南北望，整个陵区的风水形势尽在眼前，昌瑞山一览全体，除定陵、昭西陵外，其余的各陵群均朝向金星，气场强大无比，仿佛宇宙能量尽聚于此，左膝盖竟未觉任何不适，仿佛被治愈了……

3-053 Map of the feng shui situation, the Eastern Qing Tombs, Zunhua City (13 June 2019)

For the first time, I climbed Jinxing (Venus) Mountain, panting and tired. Looking from the south to the north, the feng shui situation of the entire mausoleum, including the Changrui Mountain, was in front of me. Except for the Ding Tomb and Zhao Western Tomb, all of the other mausoleums are facing Venus Mountain, creating an extremely powerful atmosphere, as if the cosmic energy were all gathered here. My left knee did not feel any discomfort, as if they were cured.

419

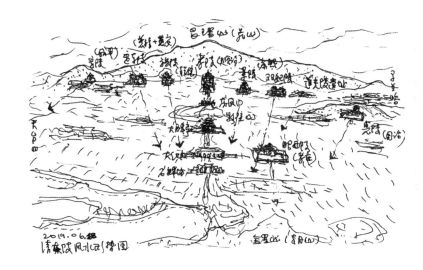

清东陵风水山势图
2019.06.28

3-054 义县及奉国寺山川形势图 （2015.11.10）

3-054 Situation map of mountains and rivers of Yi County and Fengguo Temple (10 November 2015)

421

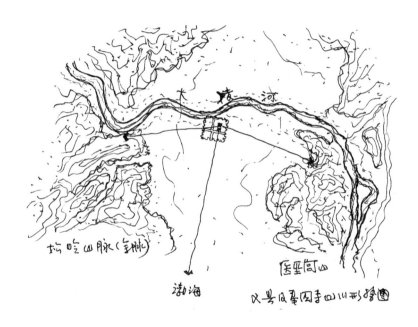

3-055 义县，城关街市示意图（民国十七年春，1928 年）
（2015.11.10）

奉国寺大殿、城中央的鼓楼、城西南角的广胜寺塔，恰好西南斜向连成一线。

3-055 Schematic plan of Town Market (Spring of 1928, the Republic of China), Yi County (10 November 2015)

The main hall of the Fengguo Temple, the Drum Tower in the city centre and the Guangsheng Temple Tower in the southwest corner of the city are located diagonally in a southwest line.

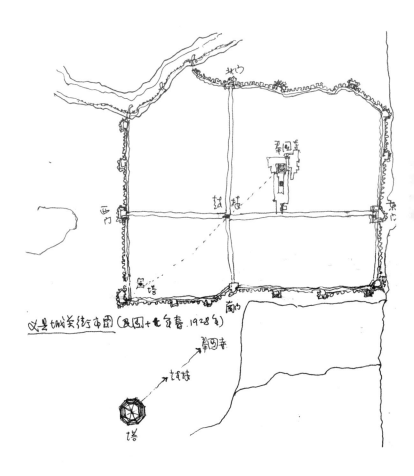

义县城关街市图（民国十七年春.1928年）

3-056 义县，奉国寺（2015.11.10）

冬至，奉国寺。站在当代的地面，千年前的古代地面沉在下方，半俯瞰视角中，奉国寺殿宇群似乎也压缩空间，沿着中轴甬道"贴"在一起。

3-056 Fengguo Temple, Yi County (10 November 2015)

I came to the Fengguo Temple during the Winter Solstice. Below the contemporary ground is the ancient one, sunk a thousand years ago. In the semi-overlooking view, the buildings of the Fengguo Temple also seem to be compressed and combined along the central axis.

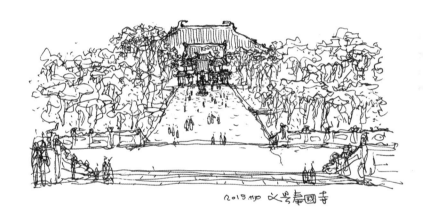

2015.11.10 义乌奉国寺

3-057 义县，奉国寺大殿，"中唐"（2015.11.10）

天地山河，弥纶宇宙，无量胜境：一个渐步揭示却突然呈现的世界。

沿着高出两侧院庭地面的甬道行进，我想到了敦煌莫高窟壁画净土经变中佛国世界的水中浮桥，沿中轴布置的群体建筑、檐廊和甬道、平台都坐落于抬升高起的台基之上，仿佛台基周侧为"莲池"环围。奉国寺的格局及其给人的体验使人犹如身处一个完整的人间净土世界。穿越层层厅堂，行步之间逐渐靠近，尺度超人的雄伟辽构大殿蓦然现于眼前。

3-057 Zhong Tang, the Grand Hall of the Fengguo Temple, Yi County (10 November 2015)

Both heaven and earth, mountain and river, the universe and the infinite realm: a world that is gradually uncovered but suddenly appears.

Walking along the paved path (*yongdao*) above the courtyard ground on both sides, I thought of the pontoon bridge floating on the water in the Buddhist world, as it appears in the pure land painting (*jingtu jingbian*) of Dunhuang Mogao Grottoes. The buildings, corridors, *yongdao* and platforms along the central axis are located on a raised base. The perimeter of the base appears to be surrounded by lotus ponds, making the Fengguo Temple a complete world of pure land. Passing through the various halls, one gradually approaches the superb grand hall of the Liao Dynasty, which is now in sight.

2015.11.10 又是春回李 中鉴唐

3-058 义县，奉国寺大雄宝殿 （2015.11.10）

七佛，手印（放下），屋架的渐变。大殿之中未施天花，排列的众佛像处于丛立、暴露、力量的纵横结构之中，特有西方教堂般的神圣升腾之感，虽然佛像其实是在结构的干扰之下处在一个不很"单纯"的背景里，但也因此更加与众生拉近了距离，所谓"人神共处"。奉国寺群体空间与莫高窟壁画中的净土经变画面空间相仿佛，后面的大殿是整个空间序列的高潮，其佛像彩画也是空间序列的不可分割部分，这是真正在人间的"理想天国"——不借助自然环境而全凭人力造成现实中的"仙境再现"，是奉国寺的最独特之处。

3-058 The Great Buddha's Hall (*Daxiongbaodian*) of the Fengguo Temple, Yi County (10 November 2015)

Seven Buddhas, handprints (putting down) and the gradual changes in the roof truss. There is no ceiling in the hall, and the Buddha statues are arrayed in vertical and horizontal structures, which are clustered and powerful. A sacred sense of the Western churches' ascension appears here. Although the background of the Buddha statues is not "pure", under the interference of the structure, they are closer to all living beings. Put another way, the so-called people and gods coexist. The group spaces in the Fengguo Temple are similar to the spaces in the pure land paintings seen in the murals of Mogao Grottoes. The rear hall is the climax of the entire spatial sequence and the Buddhist statues and paintings are an inseparable part of it. This is truly an ideal heaven in the real world. The unique aspect of the Fengguo Temple is the "representation of a fairyland" made by human beings without the help of the nature.

· 揽舍天地
· 慈满山河
· 弥纶宇宙
· 无量胜境

2015.11.10上午
义之堂四寺大雄宝殿
七(佛.手印(如下)
(屋架的清妙)

3-059 蓟县，独乐寺，山门 （2016.03.05）
　　殿宇般的山门。

3-059 Gate (*shanmen*) of the Dule Temple, Ji County (5 March 2016)
　　A hall-like gate.

2016.03.05　龙泉寺山门

3-060 蓟县，独乐寺，山门内望观音阁（2015.11.14）

过白。三层的观音阁是对观音像的围裹。建筑、人与佛像的尺度对比与呼应。观音之阁，具足圆成。"熟悉敦煌壁画中净土图者，若骤见此阁，必疑身之已入西方极乐世界矣。"（梁思成，独乐寺专号，中国营造学社汇刊，1944，3(2)，p11）独乐寺仅凭现况一门一阁，即足以呈现一个佛国理想空间。

3-060 Looking at the Guanyin Pavilion from the gate (*shanmen*), Dule Temple, Ji County (14 November 2015)

A framed scene (*guobai*). The Guanyin statue is surrounded by the three-storey Guanyin Pavilion. The scale of the architecture, people and Buddha contrast and echo with each other. A board with four Chinese characters written on it (*ju zu yuan cheng*, literally "great success") is hung in the Guanyin Pavilion. The architectural historian Liang Sicheng wrote that for anyone familiar with the pure land paintings in the Dunhuang murals, if they see this pavilion, they will think that they have entered Western Paradise ("Special issue on the Dule Temple", *The Bulletin of the Society for the Research in Chinese Architecture*,1944, 3(2), p. 11). Based on the current situation (with only a gate and a pavilion), the Dule Temple is able to present ideal Buddhist space.

具足圆成

观音之阁

2016.11.16 西山门溪观音阁 北浮寺

3-061 蓟县，独乐寺，观音阁平座层观音视线示意
（2015.11.14）

　　观音（北）之眼与朝拜者（人）、山门、城市、白塔（南）的视野连线，仿佛观音凝视众生。过白、尺度、视线；人、神，众生。

3-061 Sight diagram of Guanyin Statue on the upper floor, Guanyin Pavilion, Dule Temple, Ji County (14 November 2015)

　　The eye of Guanyin (north) is connected with the worshippers (people), the gate, the city and the white tower (south), as if Guanyin stares at all beings.

435

塔　　　山门　　　观音殿)

2015.11.14.

3-062 蓟县，独乐寺（之一）(2019.03.03)

观音阁、古树，山门，蓟城，白塔。

3-062 Dule Temple, Ji County (1) (3 March 2019)

The Guanyin Pavilion, ancient trees, gate, Ji city, white tower.

独特寺 2019.3.3.

3-063 蓟县，独乐寺（之二）(2019.03.03)

观音阁、古树，山门，蓟城，白塔。

3-063 Dule Temple, Ji County (2) (3 March 2019)

The Guanyin Pavilion, ancient trees, gate, Ji city, white tower.

独乐寺 2019. 03. 03.

3-064 五台山，佛光寺组群剖面 （2017.11.06）

东大殿位于几组台地中最高的斩山形成的平台之上，坐东面西，俯瞰下方的庭院、田野和远山、夕阳。

2017 年 11 月 5—6 日的这次再访佛光寺，使我有一种类似"顿悟"的强烈体验，佛光寺几乎可以成为一直在探寻的"现实理想空间"营造范式的实体样本，由此获得佛光寺五点启示：一、选址和方位决定圣境／胜景；二、由全凭人力转为交互自然；三、建筑意匠强化深远空间；四、漫游与沉浸并重之空间叙事；五、由竖向崇高转为平向深远。

3-064 Section of the Foguang Temple buildings, Wutai Mountain (6 November 2017)

The East Hall is located on the highest platform of several sets of terraces formed by cutting into the mountains. It faces the west, overlooking the courtyard, with fields, distant mountains and the setting sun.

Revisiting the Foguang Temple on 5–6 November 2017 was a strong experience for me, like an epiphany. The Foguang Temple can almost become a physical example of the paradigm of building "realistic ideal space" that has been explored. I propose five points of inspiration, learned from the Foguang Temple: (1) site selection and orientation determine the poetic scenery; (2) transformation from human imposition over nature to becoming more interactive with nature; (3) architectural conceptions strengthen spatial depth; (4) spatial narrative of wandering and immersion with equal importance; and (5) transformations from the vertical sublime to the horizontal deep distance.

远山

2017.11.06. 佛光寺山群剖面

3-065 五台山，佛光寺东大殿内西望 （2017.11.05）

佛坛前面的前廊，打开的大门，双松，经幢，下方建筑屋顶，树丛，远山，空间层层推远。

3-065 Looking west from the East Hall of the Foguang Temple,Wutai Mountain (5 November 2017)

The front porch of the Buddhist altar, the opening door, the double pines, the Buddhist stone pillar, the buildings' roofs below, the trees and the distant mountains produce multi-layered and far-reaching space.

2017.11.05   佛光寺东大殿内西望

3-066 五台山，佛光寺，东大殿前平台面西望庭院和群山、夕阳（2017.11.05）

"山如佛光"。

3-066 Looking west to the courtyard and mountains at sunset on the front platform of the East Hall, Foguang Temple, Wutai Mountain (5 November 2017)

The mountains are shining under the sunset, as if they were lit by the Buddha's light.

2017. 11. 05   "山村的阳光"

3-067 五台山，佛光寺后山远眺 （2017.11.06）

三年后再于东大殿后山西眺寺群、原野、远山，并绘图。西方群山前，原野被两侧的河道沟壑切成"半岛"，并向近处延伸上来，形成佛光寺所在的多级台地，也被两侧冲沟包裹。当人力的营造与大地自然的造化相得益彰，就会产生文明的杰作。

3-067 Overlooking from the back hill of the Foguang Temple, Wutai Mountain (6 November 2017)

Three years later, I once again overlook the temple buildings, wilderness and the distant mountains in the west while standing on the back hill of the East Hall, and I draw a sketch. In front of the western mountains, the distant wilderness is cut into a "peninsula" by rivers and gullies on both sides. The "peninsula" extends to the vicinity, forming multi-level platforms where the Foguang Temple is located. It is also surrounded by gullies on both sides. When the creations of man and the natural creations of the earth complement each other, a masterpiece of civilisation is produced.

2014·11.06 佛光寺后山远眺

3-068 五台山，摹卢绳先生佛光寺全景（2017.11.06）

致敬先师。

3-068 Imitating Lu Sheng's panoramic drawing of the Foguang Temple, Wutai Mountain (6 November 2017)

Paying tribute to Lu Sheng, architecture professor of the older generation at Tianjin University.

2017.11.06

蒋子琅先生佛光寺之景

3-069 北京，故宫，东路甬道（2019.01.17）

　　高大红色宫墙与常规尺度青砖建筑院落群的并置，甬道尽端是金碧辉煌的宫殿和角楼。

3-069 A paved path leading to a main hall in the eastern part of the Forbidden City, Beijing (17 January 2019)

　　The tall, red palace wall is juxtaposed with conventionally scaled grey-brick buildings and courtyards. At the end of the path, there are magnificent palaces and a turret.

2019.01.17 故宫东路甬道

3-070 北京，故宫，宁寿宫花园（乾隆花园）总平面图
（2019.06.01）

高大宫墙背后的庭园群。五个院落／庭园以一条中间折动的轴线串联，各具主题特征。北端是统领性的符望阁庭园和倦勤斋庭园。

宁寿，倦勤，禊赏，符望——参加丁垚老师一年级空间认知课之意外收获：冯建逵先生图绘乾隆花园，20 世纪 50 年代卢绳先生带天津大学建筑系学生测绘符望阁剖面图，卢绳先生 1957 年日记测绘草图。

3-070 General plan of Ningshou Palace Garden (Qianlong Garden), the Forbidden City, Beijing (1 June 2019)

The courtyards are behind the tall palace wall. The five courtyards/gardens, each with a thematic feature, are connected by a central, twisted axis. At the north end, the Fuwang Pavilion courtyard and the Juanqinzhai Hall garden are dominant.

Peaceful and long life (*Ningshou*), relax (*juanqin*), appreciation (*xishang*) and overlooking (*fuwang*). Unexpected gains during the space cognition course led by Ding Yao include: Feng Jiankui's drawing of the Qianlong Garden, a 1950s section drawing of the Fuwang Pavilion by architecture students and their tutor Lu Sheng of Tianjin University and Lu Sheng's 1957 surveying and mapping sketches and diary.

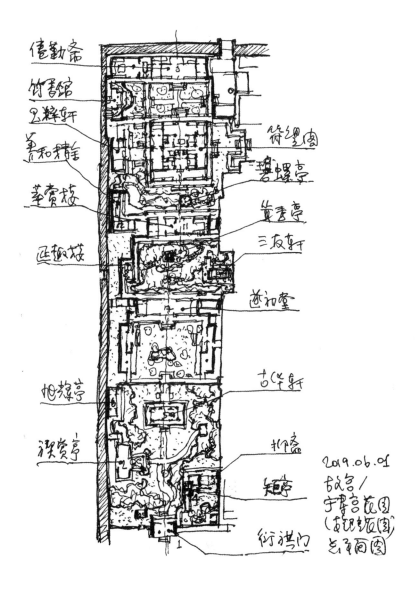

451

佳勤斋

竹香馆

玉粹轩

养和精舍

萃赏楼

延趣楼

旭辉亭

撷芳亭

符望阁

碧螺亭

耸秀亭

三友轩

遂初堂

古华轩

柳亭

矩亭

衍祺门

2019.06.05
故宫/
宁寿宫花园
(乾隆花园)
总平面图

3-071 北京，故宫，宁寿宫花园（乾隆花园），符望阁平面图和剖面图（2019.01.17）

庞大宫城之中一座小小的貌似对称严整的楼阁，且全凭人力营造，其空间体验与高山深林之中的佛光寺建筑群相比，却有异曲同工之效。陡峭而精巧的楼梯连接上下，符望阁每层空间、功能、尺度均有极大不同，犹如叠摞起来的多维世界：一层充分运用"装折"手段，制造令人惊奇的空间和尺度变化，犹如"迷楼"；北侧一、二层之间的小夹层——乾隆"仙楼"，是个神秘而有趣的存在，是经历下方空间"迷幻"之后的静息之处。

3-071 Plan and section of the Fuwang Pavilion, Ningshou Palace Garden (Qianlong Garden), the Forbidden City, Beijing (17 January 2019)

Within the huge Forbidden City, a small, seemingly rigorously symmetrical pavilion was man-made, but the space experience it creates is similar to the Foguang Temple within the deep forests and mountains. Steep and intricate staircases connect the upper and lower floors. The space, function and scale of each floor of the Fuwang Pavilion are very different, like a multi-dimensional world that is stacked up. By fully using the method of joinery work (*zhuangzhe*) the first floor to produce amazing variations of space and scale so that it appears to be a "labyrinthine building". The Qianlong *xianlou* on the north side is the mezzanine in between the ground and first floor, which is mysterious and interesting. It is a resting place after experiencing the "psychedelic" space below.

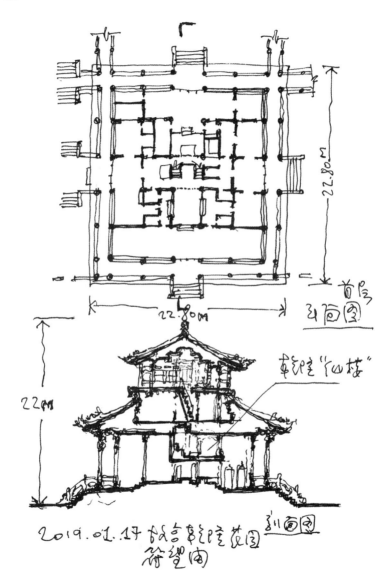

首层
平面图

22.80m

22.80m

仿建"仙楼"

22m

剖面图

2019.01.17 故宫乾隆花园
符望阁

3-072 北京，故宫，宁寿宫花园（乾隆花园），符望阁首层北侧入口厅正面速写（2019.01.17）

正中分上下两个夹层，下面似为静处之所，正襟危坐，"静寄参心得，虚怀契道存"（座后联）；上层为乾隆"仙楼"，空荡自在，"四壁图书鉴今古，一庭花木验农桑"（"仙楼"外联）。南侧正入口厅联："即事畅天倪知仁同乐，会心成静寄远近咸宜。"

3-072 Sketch of the Fuwang Pavilion's front entrance (north side) on the ground floor, Ningshou Palace Garden (Qianlong Garden), the Forbidden City, Beijing (17 January 2019)

The central hall is divided into two levels in the middle. The lower level is a quiet and formal place, while the upper one is *xianlou*, empty and comfortable.

四壁图书鉴赏古
一庭花木好栽培

新年无似得
随怀契真在

眼看明天便知仁同年
金心成新年道道减道

3-073
北京，故宫，宁寿宫花园（乾隆花园），符望阁首层北侧入口厅（2019.01.17）

"纵目恰宜遇寥朗，讬怀又喜得清舒。"（乾隆"仙楼"内联）

符望阁二层为书房，尺度适宜，是内向而封闭的私人空间；三层则是宽阔敞亮，屏风龙座坐北朝南，四周开窗并设环廊可登临望景，体现"符望"之意。这不是一个普通的景观楼阁，而是乾隆皇帝在威严宏大的紫禁城中留给自己的一处私人住宅，充满身体感、空间感、人文感、叙事感、当下感，几乎可称是能与帕拉迪奥别墅相媲美的东方"当代"建筑设计杰作。

3-073 The north entrance hall of the ground floor of the Fuwang Pavilion, Ningshou Palace Garden (Qianlong Garden), the Forbidden City, Beijing (17 January 2019)

The second floor is a study room with an appropriate scale. It is an inward and closed private space. On the wide and bright third floor, the emperor sits on the seat in front of the screen while facing south. The space, surrounded by circular corridors and windows that open, offers a place to climb up and view the surrounding landscape, embodying the meaning of 'Fuwang'. This is not an ordinary landscape pavilion, but a private residence in the majestic Forbidden City, once owned by the Emperor Qianlong. Full of humanity, bodily, spatial, narrative and present senses, it is an oriental masterpiece of "contemporary" architecture comparable to the Palladian Villa.

纵目始宜逢豁朗
托怀又喜得清舒
故宫宁寿宫花园
符望阁／乾隆

2019.01.坪

3-074 日本岐阜县高山市，吉岛家住宅（日下部家住宅与其有类似之处）：被筱原一男称为"这才是建筑"的建筑（2017.12.17）

框架木梁与格子纸障——宏大高拔的象征性空间与平易深远的日常性空间暧昧并置；出乎意料的叙事——开门见山、惊心动魄，而后温语绵绵、层叠不尽；喧嚣与静谧共存——像是一个极小的城市，有"街巷"或者"广场"一样的空间，和宅居、园庭既融合，又隔分。

3-074 Yoshijima Residence (the Kusakabe Heritage House is similar to it in part), Takayama City, Gifu Prefecture, Japan. : Kazuo Shinohara called it "true architecture" (17 December 2017)

The wooden frame and beams are complemented by plaid paper barriers. It is an ambiguous juxtaposition between grand and high symbolic space and plain and deep daily space. This is an unexpected narrative, straightforward, thrilling and then lingering and cascading. Chaos and tranquillity coexist, like an extremely small city, with "streets" or "square space", both integrated with and separated from houses and gardens.

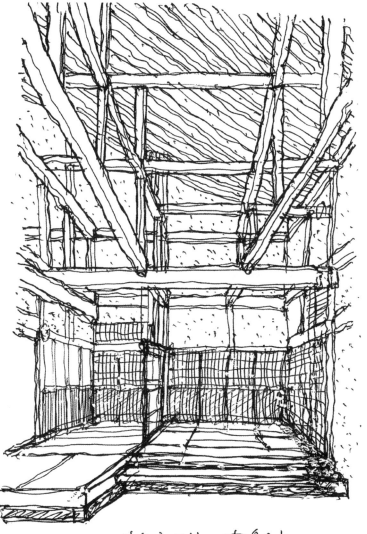

2017.12.17 吉岛家伦笔，岐阜县高山市，1907年

3-075 坂本一成，东京多摩，Hut AO（2015.08.28）

与坂本一成自宅 House SA 相比，这个为女儿设计的私宅像是缩小版的"螺旋"之家。

坂本先生的教导：

形式的危险和微妙平衡。

怎样获得自由——文化的自由：个人思考随对当下社会的观察及批判而变；空间的自由：内外的连续性＝边界和限定的消除；建筑师的自由："马马虎虎"＝放松和柔软。

3-075 Hut AO designed by Kazunari Sakamoto, Tama, Tokyo (28 August 2015)

Compared with Sakamoto's House SA, this private house designed for his daughter is like a miniature version of the 'spiral' house.

Sakamoto's teaching: the danger and subtle balance of form.

How to get: (1) the freedom of culture: personal thinking changes with the observation and criticism on the present society; (2) the freedom of space: the continuity of inside and outside equals the elimination of boundaries and limits; (3) the freedom of architects: carelessness equals relaxation and softness.

461

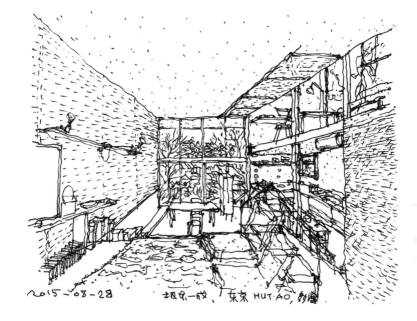

2015-08-28　城见一成，东京 HUT AO, 外观

3-076 杰弗里·巴瓦，科伦坡，希马玛拉卡寺，"水中庙"

（2015.01.17）

水中小世界：由桥涉水而入，居中大树，大小庙宇左右两分，在庭院中自由分布而不求对称。

斯里兰卡的杰弗里·巴瓦，或许可以成为对我们最具启示性的、世界级的亚洲建筑师。他所营造的那一个个大小世界，人造物和自然物像恋爱的男女，无分主次、相反相融。只有在现场亲身体验，才能感受得到那种介于尘俗与雅意之间的动人力量，无论是建筑师还是景观设计师，艺术家、开发商，日常的使用者还是仅仅到此一游的人。

3-076 Seema Malaka Temple designed by Geoffrey Bawa, 'Temple in Water', Colombo (17 January 2015)

A small world in the water: walking over the bridge, by the big tree in the centre, the large and small temples that are separated and freely distributed in the courtyard without symmetry.

Geoffrey Bawa was born in Sri Lanka. As an inspiring world-class Asian architect, he might deserve more of our attention than he has had. In the large and small worlds he had created, man-made and natural objects are integrated with equal weight, both contrast and complementary, like men and women in love. Only when they are experienced on site can the inspiring force between the mundane and the elegant be felt, whether the viewer is an architect or a landscape designer, an artist or a developer, a daily user or just a visitor.

2015-01-17

Beira Lake.
Seema Malaka Temple
Colombo (1976-78)

有了绿色大树，
"以中画"

3-077 杰弗里·巴瓦，科伦坡，希马玛拉卡寺，"水中庙"，大庙（2015.01.17）

混凝土基座平台上神奇的木构殿宇，宏伟与雅致并存，内部空间自由、静谧、安逸而日常，人们席地而坐于通透的庙堂之中，是空间的主人。

3-077 Seema Malaka Temple designed by Geoffrey Bawa, 'Temple in Water', Colombo (17 January 2015)

In the magical wooden structure built on the concrete platform, grandness and elegance co-exist. The interior space is free, quiet, comfortable and common. People sitting within the transparent temple is the master of the space.

2015-01-17 水中庙
Seema Malaka Temple
Colombo (1976-78)

3-078 杰弗里·巴瓦，科伦坡，巴瓦工作室 (2015.01.07)

于垂直建筑空间主轴的龛座中坐而静观，水池、行人、庭院、塑像、对坐者，扁平压缩而"深远"的空间与景象。

3-078 Bawa Studio/House for Dr Bartholomeusz designed by Geoffrey Bawa, Colombo (7 January 2015)

Sitting in the niche, perpendicular to the building's main spatial axis, I quietly observe the pool, pedestrians, courtyards, statues and the person sitting opposite. I am experiencing the flat and compressed scene and the "distant" space.

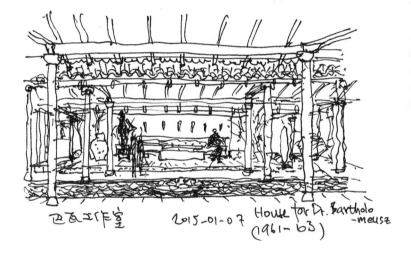

卫生工作室　　　2015-01-07　House for Dr. Bartholo
　　　　　　　　　　　　　　　　　(1961- 63)　　-meusz

3-079 杰弗里·巴瓦，科伦坡，巴瓦工作室 /Paradise Road Galleries 咖啡馆 (2015.01.25)

　　斯里兰卡之行的起点和终点。壁龛之"座"与对景的三重空间：水池、通廊与庭院（雕塑、对座）。

3-079 Bawa Studio/Paradise Road Galleries Cafe designed by Geoffrey Bawa, Colombo (25 January 2015)

　　The start and end points of the trip to Sri Lanka. There are triple-layered space opposite the seats in the niche: the pool, the corridor and the courtyard (sculpture, opposite seat).

巴乐之北画堂
Paradise Road
Galleries
(०१०००५५५६)

2015-01-25
Srilanka Trip started and
ended Point.

3-080 杰弗里·巴瓦，科伦坡，巴瓦工作室入口庭院 /No.5 Alfed Place (2015.01.17)

大门的摆线拱过白。

260 Bawa Studio Entrance Courtyard / No.5 Alfed Place designed by Geoffrey Bawa, Colombo (17 January 2015)

The framed scene (*guobai*) created by the cycloidal arch of the gate.

No.5 Alfred Place, Colombo 3　巴瓦工作室入口庭院

(1961~63)　　　　　　　2015-01-1半

3-081 杰弗里·巴瓦，科伦坡，新议会大厦（2015.01.17）

如同一组巨大的"水中庙"。大小尺度组合，木石上下组合（上木下石）。对称与不对称结合的宇宙观。

3-081 New Parliament Building designed by Geoffrey Bawa, Colombo (17 January 2015)

This is like a group of huge "temples in water", with a combination of large and small scales. The upper part is wood and the lower part is stone. The cosmology is represented through a combination of symmetry and asymmetry.

新议会大厦   New Parliament
(1979-82)   2015-01-17

3-082 杰弗里·巴瓦，科伦坡，33号街巴瓦自宅楼梯间
（2015.01.18）

"弯曲"一体，如镂空雕塑般的形式与空间之美。

3-082 Stairwell of the 33rd Lane Bawa House designed by Geoffrey Bawa, Colombo (18 January 2015)

The bending and curved staircase showcases the beauty of form and space similar to a hollowed out sculpture.

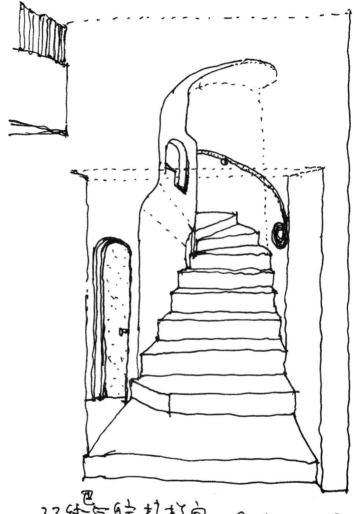

33街气脆楼梯间
（1956~98）　　2015-01-18

3-083 杰弗里·巴瓦，科伦坡，33号街巴瓦自宅入口廊道

（2015.01.18）

小庭院们作为"负形"的建筑体量，塑造建筑内部虚实互映的深远空间。

3-083 Entrance corridor of the 33rd Lane Bawa House designed by Geoffrey Bawa, Colombo (18 January 2015).

As "negative" building volume, the small courtyards shape the far-reaching space inside the building.

Pradeep Jayawardene House, Mirissa   2015-01-23
(Red cliff Hill Villa) (1997~98)   赤壁之家
In front of the Weligama Bay, South Coast

3-085 杰弗里·巴瓦，本托塔，巴瓦哥哥庄园（2015.01.21）

岁月沧桑包浆，几乎黝黑的石质塑像，"鸡鸡"上面居然是一丛鲜绿的嫩草，好一处特别的"人工与自然的互成"。

3-085 The home and Garden of Bevis Bawa designed by Geoffrey Bawa, Bentota（21 January 2015）

After years of oxidation, the stone statue has become almost dark. A cluster of bright green grass grows on the statue's penis. What a special integration of the artificial and the natural!

481

The Home & Garden
of Bevis Bawa,
Rabawila Village, Dehi
12070 Sri Lanka - wala

2015 - 01 - 21
巴瓦哥之生園

3-086 杰弗里·巴瓦，本托塔，卢努甘卡庄园印象（2015.01.26）

将人造物——一个大水缸置入自然（林边坡地），作为视线焦点，与远山、寺庙、草地、近树共同成为可观想的对景。"用一个缸教化一个整体景观"——以自然开始，以一个"自然"要素结束。

3-086 Impression of Lunuganga Garden designed by Geoffrey Bawa, Bentota (26 January 2015)

A large man-made water tank was embedded into nature (a slope near a forest). As a sight focus, together with the distant mountains, temples, grasslands and nearby trees, it becomes part of the mutually corresponding scenery for viewing and meditating. Using a tank to civilise a holistic landscape—starting with nature and ending with a "natural" element.

483

卢努甘卡庄园印象   2015-01-26
Lunuganga Garden (GB's country estate), Bentota
(1948~98)卢努甘卡, 乌拉图博.

3-087 杰弗里·巴瓦，丹布拉，坎达拉玛遗产酒店，"人工绿崖"（2015.01.20）

"非建筑感"的平屋顶＋种植屋面，混凝土柱＋木构架爬植，制造消隐的"人工山体"。在显藏之间，建筑与山崖融为一体。

3-087 Artificial green cliff, Kandalama Heritage Hotel designed by Geoffrey Bawa, Dambulla (20 January 2015)

The "non-architectural" flat, planted roof, concrete columns and climbing plants on wooden frames create a hidden artificial mountain. Oscillating between the visible and the hidden, the building and cliff are combined into one.

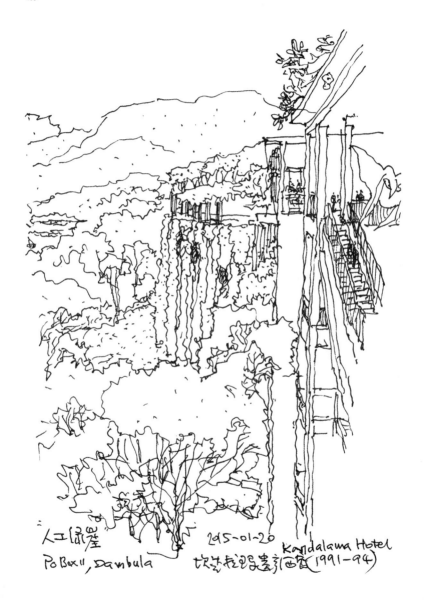

"人工园罢"

Po Box 11, Dambula

2015-01-20 Kandalawa Hotel

坎兹拉别墅引面窝(1991-94)

3-088 杰弗里·巴瓦，丹布拉，坎达拉玛遗产酒店，"无限泳池"（2015.01.20）

巴瓦在商业酒店中首创了这一手法。沉浸于泳池中遥望大湖远山，在透视原理作用下，近水与远水在视觉上融为一体。

3-088 Infinity pool, Kandalama Heritage Hotel designed by Geoffrey Bawa, Dambulla (20 January 2015)

Bawa pioneered the infinity pool approach in commercial hotels. Immersed in the pool and overlooking the great lake and distant mountains, one may find that the nearby water and the distant water are visually integrated under the principle of perspective.

坑连接江马造引酒店 天眼湖4史 2015-01-20

3-089 杰弗里·巴瓦，丹布拉，坎达拉玛遗产酒店，"无限泳池"（2015.01.20）

聪明的营造法。

3-089 Infinity pool, Kandalama Heritage Hotel designed by Geoffrey Bawa, Dambulla (20 January 2015)

A smart design and construction method.

もたせはら浄水場
(1991〜94)

2015-01-20

3-090 杰弗里·巴瓦，丹布拉，坎达拉玛遗产酒店，入口 （2015.01.20）

　　圆柱，平坦而深邃的雨篷，甚至灯光，把酒店入口从外部山林中勾勒出来。

3-090 Entrance of the Kandalama Heritage Hotel designed by Geoffrey Bawa, Dambulla (20 January 2015)

　　The column, the flat and deep canopy, and even the light, outline the hotel entrance from the mountain and forest outside.

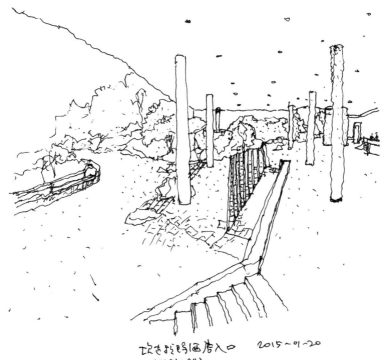

文化およう鳴海店入口
（1991〜94）　　2015〜01〜20

3-091 杰弗里·巴瓦，本托塔，海滩酒店（2015.01.22）

叠加，悬挑，消失的标准层。对应巴瓦草图的建筑构成：水面、地基、底座、基座、地面、墙柱、屋顶——巴瓦建筑中的自然之物——水、地、石、木、天，与建筑的密切互动。

3-091 Beach Hotel designed by Geoffrey Bawa, Bentota (22 January 2015)

The superimposed, cantilevered and disappearing standard floors. The building elements corresponding to Bawa's sketch: water, foundation, base, pedestal, ground, wall/pillar and roof. The natural elements in Bawa's architecture: water, earth, stone, wood and heaven, and each one has close interaction with the building.

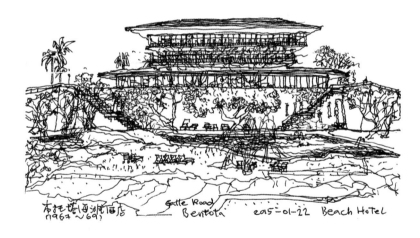

斯里兰卡海滨酒店
(1967～69)

Galle Road
Bentota   2a5-01-22 Beach Hotel

3-092 杰弗里·巴瓦，本托塔，海滩酒店，庭院（2015.01.21）

通透的凸出檐廊环绕着方池，池中方形的 "树之岛" 自由分布，巨大的树冠悬挑于水面之上，几乎充满了庭院。被抽象化、人工化和形式化，而又真实存在的自然。

3-092 Courtyard of the Beach Hotel, Bentota (21 January 2015)

The transparent protruding porch surrounds the square pool, and the square "tree island" in the pool is freely distributed. The huge canopy overhangs the water and almost fills the courtyard. The scene is abstract, artificial and formal, but still an truly existing nature.

海滩海岸度假院
Galle Road, Bentota

2015-01-21

Bentota Beach Hotel,
(1967~69)

3-093 杰弗里·巴瓦，本托塔，海滩酒店，水院小岛细部
（2015.01.22）

池（岸），水（口），草，树，罐子。

3-093 Details of a small square island, water courtyard, Beach Hotel, Bentota (22 January 2015)

Pool (shore), water (mouth), grass, trees, jars.

沙网起画后水笑小酋细部
Dean Hotel, Bentota (1967~69)   2015-01-22

3-094 杰弗里·巴瓦，本托塔，海滩酒店，大堂 （2015.01.22）
空间与庭院。

3-094 Lobby of the Beach Hotel, Bentota (22 January 2015)
Space and courtyard.

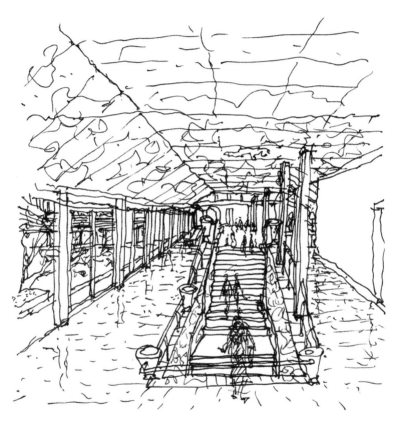

海边钊酒店 大堂
Beach Hotel, Lobby
(in 1 7~69)

2015-01-22
Bentota

3-095 杰弗里·巴瓦，本托塔，Avani 酒店，前台（2015.01.22）

空间与庭院。

3-095 Reception of Avani Hotel designed by Geoffrey Bawa, Bentota
(22 January 2015)

Space and courtyard.

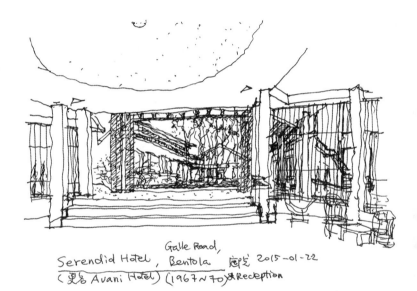

Galle Road,
Serendid Hotel, Bentola 屋塹 2015-01-22
(更8 Avani Hotel) (1967~70)※Recdeption

3-096 杰弗里·巴瓦，阿洪加拉，遗产酒店，大堂（2015.01.24）
　　不同材质、颜色构成的"柱林"，建筑结构（柱梁）对自然（树木）的仿造。

3-096 Lobby of Triton Hotel designed by Geoffrey Bawa, Ahungalla (24 January 2015)
　　The "forest" of columns is composed of different materials and colours. The building structures (columns and beams) imitate nature (trees).

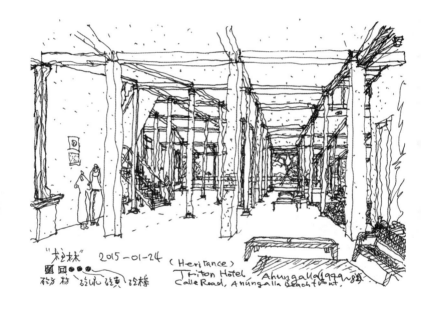

"栏林" 2015-01-24 (Heritance)
Triton Hotel   Ahungalla(1999~8新)
Calle Road, Anúngalla Beachfront.

筋材 弦瓦 弦黄 弦棕

3-097 (1-2) 杰弗里·巴瓦， 阿洪加拉，遗产酒店，柱林 + 水池 （2015.01.24—25）

柱林中的水池，池水中的树岛。

3-097 (1-2) "Forest" of columns and the pool, Triton Hotel designed by Geoffrey Bawa, Ahungalla (24, 25 January 2015)

The pool in the "forest" of columns and the island of trees in the pool.

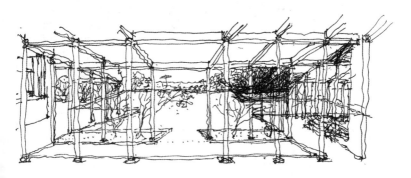

柱林 + 水（池）
Triton Hotel, Ahungalla
(1979~81)
(Heritance)

2015-01-24
阿洪加拉遗产酒店

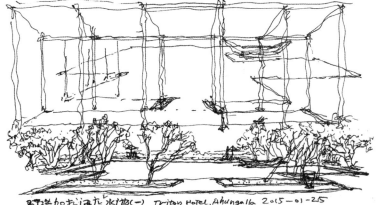

阿珀加拉酒店 水池(一) Triton Hotel, Ahungalla 2015-01-25
堂 (1979~81) (Heritance)

3-098 杰弗里·巴瓦，阿洪加拉，遗产酒店，大树 + 水池

（2015.01.24）

几何蜿蜒的水池与悬挑其上的大树，人仿佛在丛林下的自然湖水中游泳。

3-098 Big tree and pool, Triton Hotel designed by Geoffrey Bawa, Ahungalla (24 January 2015)

The pool is geometrically extending and there is a big tree overhanging it. People appear to be swimming in a natural lake in the jungle.

大榕 + 水(山也(三)
(Heritance)
Triton Hotel, Ahungalla

2015-01-24
阿蓝加对这远引画台

3-099 杰弗里·巴瓦，瓦杜沃，碧水酒店，大堂（2015.01.25）
水成为建筑秩序（人工）中的一部分。

3-099 Lobby of the Blue Water Hotel designed by Geoffrey Bawa,
Wadduwa (25 January 2015)
Water becomes part of the architectural order (artificial).

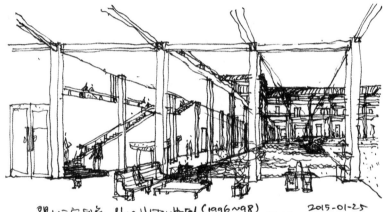

碧水酒店/玉印象 Blue Water Hotel (1996~98)　　　2015-01-25
Thalpitiya, Wadduwa Beachfront, Wadduwa

3-100 杰弗里·巴瓦，加勒，灯塔酒店，入口大坡地速写 （2015.01.24）

人工化的坡地。

3-100 Sketch of the entrance to the slope, Lighthouse Hotel designed by Geoffrey Bawa, Galle (24 January 2015)

Man-made slopes.

村庄长峪店
古地道县 Lighthouse Hotel, Dadalla, Galle 2015-01-24
1995～97

3-101 杰弗里·巴瓦，加勒，灯塔酒店，檐廊与庭院
（2015.01.24）

将自然圈（植）入建筑空间，使之成为空间、形式不可或缺的一部分。

3-101 Corridor and courtyard, Lighthouse Hotel designed by Geoffrey Bawa, Galle (24 January 2015)

The nature embedded (planted) in the building space becomes an indispensable part of the space and form.

灯塔西店庭院
Lighthouse Hotel
(1995-97)

2015-01-24

3-102 杰弗里·巴瓦，Piliyandala，综合教育学院（2015.01.21）
几何控制的线型建筑"穿行"在山林间，"自然而然"的营造。

3-102 Institute for Integral Education designed by Geoffrey Bawa,
Piliyandala (21 January 2015)

The geometrically controlled linear building is located freely in
the mountains and forests. It is a "natural" design and construction.

Insitute for Integral Education
(靜坐教育科院) (1978~81)

2015-01-21

3-103 杰弗里·巴瓦，马特勒，卢哈纳大学，总平面图（2015.02.10）

结合地形，居山引水的营造；道路串联整个校区，局部空间放大；无主导轴线，以 3 米 × 3 米（平面）＋1.5 米（高度）坐标网格控制；处理高差的手段：台阶、台地、基座、建筑插入 / 直落山体，不同标高的入口和建筑体量；建筑布置：不同类型的院落——院子作为设计的主体，院子是树的主人，人是房子 / 实体的主人；岩石与水：岩石防洪，水用于生活、景观和调节气候。

3-103 General plan of University of Ruhuna designed by Geoffrey Bawa, Matara (10 February 2015)

The university was built on the mountain, combined with the topography and engaged with the lake. The entire campus, connected by roads with partially enlarged space, is not dominated by its axis but controlled by a coordinate grid that is 3 m × 3 m (in plan) and 1.5 m (in height). The architect used steps, terraces and pedestals to deal with the height differences. Part of the building was inserted into the mountain or just fell on the mountain, resulting in various entrances on different levels and different building volumes. Different types of courtyards, as the focus of the design, are involved in the overall layout. The courtyard is the owner of the trees, whereas people are the owners of the house/entity. Rock and water: rock is used to control flooding; water is used for life, landscape and climate regulation.

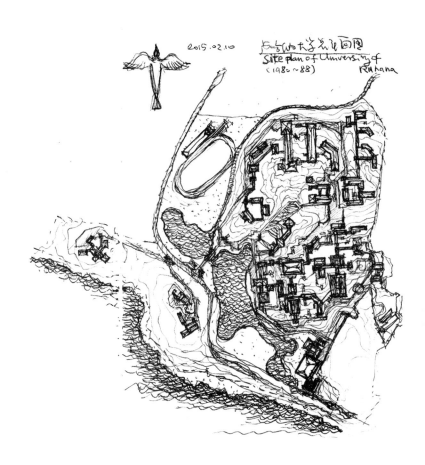

2015.02.10

斯里兰卡鲁哈纳大学总平面图
Site plan of University of
(1980~88)   Ruhana

3-104 杰弗里·巴瓦，马特勒，卢哈纳大学，植物学系教学区 （2015.02.04）

顺应山坡地形的营造：跌廊、建筑和半开放庭院。下砼上木的结构组合：混凝土柱梁＋木梁／椽坡屋顶。

3-104 The teaching area of the botanical department, University of Ruhuna designed by Geoffrey Bawa, Matara (4 February 2015)

The buildings are built on the hillside terrain: the falling corridors, buildings and semi-open courtyards. The structures combine concrete (lower) and wood (upper): concrete columns and beams plus wooden beams and a sloping roof.

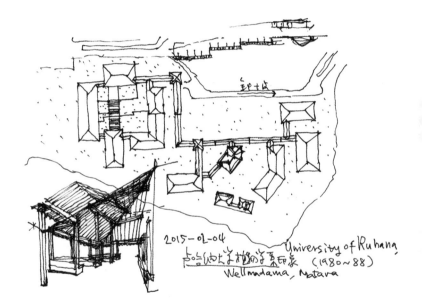

2015-01-04

ロザリンドの大学　設計者の建築印象　University of Ruhana (1980~88)
Wellmadama, Matara

3-105 杰弗里·巴瓦，马特勒，卢哈纳大学，数学系中厅

（2015.02.10）

　　董豫赣称之为"百柱厅"，一个高大的开放性公共空间（两侧是相对封闭的大阶梯教室）——"柱林"再现，犹如穿越一个林中空地。

3-105 Central hall, department of mathematics, University of Ruhuna designed by Geoffrey Bawa, Matara (10 February 2015)

　　Dong Yugan called this the Hundred Column Hall, which is tall and open (with relatively closed large lecture rooms on both sides), where people can walk through the reappearing "column forest" as if they are crossing a forest.

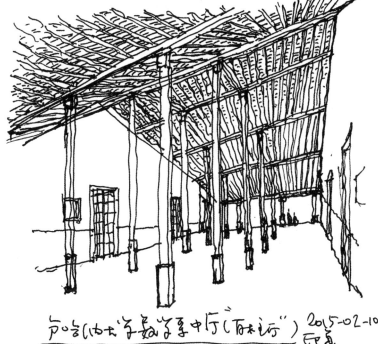

户营（地书岁数学季中厅（百未诗"）2015-02-10
印象

University of Ruhana (1980~88)
Wellmadama, Matara.

3-106 科伦坡，国家博物馆庭院中的大树（2015.01.18）

独木成林，纷繁而有序，仿佛无限延伸的深远空间，犹如这个世界。

3-106 A big tree in the National Museum's courtyard, Colombo (18 January 2015)

The lone tree becomes a forest, diverse and orderly, as if the far-reaching space extends infinitely, just like this world.

2015-01-18
斯里兰卡国家博物馆 大树 National Museum Colombo

3-107 科伦坡，Kelaniya 庙亭（2015.01.18）

石作台基上的木作庙亭。两侧第一步特别形状的石阶，表达抬升的开始，即将抵达另一个世（境）界。最简的营造。

3-107 Kelaniya Temple Pavilion, Colombo (18 January 2015)

The wooden temple pavilion on its stone base. The first stone step on the two sides, with a special shape suggests the beginning of rising and the imminent arrival of another world (realm). It is the simplest creation.

2015 - 01 - 18
Kelaniya Temple
Pavilon

3-108 丹布拉，石窟寺 （2015.01.20）
廊中起伏深远的空间。

3-108 Cave Temple, Dambulla (20 January 2015)
　　The undulating and far-reaching space in the corridor.

丹布拉石窟寺
(公元前一世纪)

2015-01-20
Dambulla

3-109 丹布拉，石窟寺（2015.01.22）

　　嵌凿于山体中的寺庙，岩石与屋顶瓦作（铜）融为一体。

3-109 Cave Temple, Dambulla (22 January 2015)

　　The temple is embedded in the mountain, and the copper roof tiles are integrated with the rock.

丹布拉石窟寺印象，Dambulla

2015-01-22

3-110 丹布拉，锡吉里耶，狮子岩顶城池宫殿遗迹（之一）

（2015.01.19）

水池＋岩石＋砖城＋树，可以抽象化的台地园林视之，座处缺亭。

3-110 Palace ruins on top of Lion Rock, Sigiriya, Dambulla (1) (19 January 2015)

The pool, rock, brick buildings and trees can be viewed as an abstract terrace garden without a pavilion at the sitting area.

狮子岩，水池＋岩石＋砖城＋树
（公元前5世纪）

2015-01-19
Sigiriya
（5世纪 B.C）

3-111 丹布拉，锡吉里耶，狮子岩顶城池宫殿遗迹（之二）
（2015.01.19）

人工城池台地遗迹与自然山川林木的相映与"对仗"。三两人群游于其间，显示出营造、使用与沉浸的主体——人（包括其中的坐而绘图者）。

3-111 Palace Ruins on top of Lion Rock, Sigiriya, Dambulla (2) (19 January 2015)

The ruins of the man-made city, with its pool and terraces, and the natural mountains and rivers, compete and contrast with each other. A few groups of people walk in between, demonstrating the main subject of creation, use and immersion: people (including the person who sits and draws).

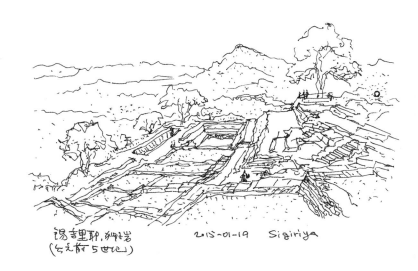

錫里耶，狮子岩
（公元前5世纪）

2015-01-19　Sigiriya

3-112 加勒，古城墙边的建筑 （2015.01.24）

城墙（遗址公园）与建筑群落并置，中间相隔带状的绿地 / 广场。

3-112 Buildings on the edge of the ancient city wall, Galle (24 January 2015)

The city wall (relics park) is juxtaposed with a group of buildings, but separated by green strips/squares in the middle.

加勒，古城墙边建筑

2015 - 01 - 24
Galle Ancient City

3-113 加勒，古城墙边的寺庙、街道、商业和民居，小雨中速写（2015.01.24）

建筑尺度与城的尺度对映，居于其中的人的多样生活。

3-113 Sketch of temples, streets, commercial buildings, and houses on the edge of the ancient city wall in the light rain, Galle (24 January 2015)

The scale of the buildings corresponds to the scale of the city, and the people who live in it have diverse lives.

加勒古城，寺庙+民居 小雨中速写　　Galle City　2015-01-24

3-114 路易斯·巴拉干，瓜达拉哈拉城，罗伯斯·里昂住宅改造（2017.01.09）

巴拉干的第一个独立作品。

实地探访巴拉干的建筑，是长久以来的心愿。巴拉干的建筑和花园，乃是一种与外部世界隔离的"家园"般心灵回归安放之地。这"家园"，与中国的"宅园"似有相通，又似有不同。如有相通处，则巴拉干早已用当代手法达到"曲折尽致""静谧胜景"之境界，值得好好鉴学；一切想象和疑问，都须前去他的现场，验证和体会。

3-114 Robles Lyon House Renovation designed by Luis Barragan, Guadalajara (9 January 2017)

This is Barragan's first independent work.

It is a long-cherished wish to visit his buildings. Barragan's buildings and gardens are an intimate and peaceful place to return to and leave the spirit, isolated from the outside world. This "homeland" seems to be both similar to and different from China's house-garden. If there are similarities, they are due to Barragan's use of contemporary methods to achieve the realms of "twists and turns", tranquillity and the poetic scenery, which is worthy of a good study. For all of these imaginations and questions, we need to go to the site of his buildings to verify and experience it.

2017.01.09. 众古老的骨彩 罗的斯老城住宅, 建于 1929

3-115, 116, 117

路易斯·巴拉干，瓜达拉哈拉城，冈泽勒斯·鲁纳住宅（2017.01.09）

　　巴拉干和筱原一男类似，也是把住宅作为"建筑"的建筑师，里昂住宅改造、鲁纳住宅、弗兰克住宅是第一时期的处女作和代表作品，其花园对应于其深受影响的布迪南德·贝克地中海风格插画，可称为"插画式花园"。令人惊异的是他才 24—27 岁（还学的是水利工程），就在短短几年间从无到有，快速发展提升，并定型出延续一生的空间模式和设计方向，这也太早熟了。不过也说明思考和实践方向定位、聚焦的重要性。

3-115, 116, 117 Gonzalez Luna House designed by Luis Barragan, Guadalajara (9 January 2017)

　　Barragan is similar to Kazuo Shinohara, because they are both architects who regard houses as architecture. The renovation of Lyon House, the Luna House, and the Frank House serve as Barragan's debut and are representative works of his early years. The garden design of these houses was influenced by Ferdinand Bac's Mediterranean-style illustration, and thus can be called the "Illustrated Gardens". What is amazing is that he was only 24 to 27 years old at the time (and studying water conservancy engineering). In just a few years he experienced rapid development and improvement, and established a space mode and design direction for a lifetime. This is too premature. However, it shows the importance of positioning and focusing on ideas and practice.

2017.01.09 飞达尔哈拉皇纳住宅，巴拉干. 1929

2017.01.09. 威迪武哈名礼 鲁迪旅馆. 巴扎干. 1929

2017.01.09 屋(四川省) . 四郎才. 1929

3-118 路易斯·巴拉干，瓜达拉哈拉城，克里斯托住宅庭院 （2017.01.09）

在巴拉干的第一时期作品中，我最喜欢和对我最有启发的是瓜达拉哈拉市的克里斯托住宅，这是巴拉干26岁时完成的市长私宅，达到惊人的成熟度：一个紧凑的城市住宅中，隔离城市喧嚣的"家园"静谧之氛围，空间节奏和"布景"变化，立体化的花园，房间一样的屋顶平台，叙事开始和结束的椭圆拱形元素。在这个100多平方米的房子里，87年前的他用直感经验把握下的"建筑性"的手法，做到了精彩的当代"宅园"/"家园"的营造，生活空间与精神空间相生互成。其后的60年时间里，他只不过是把这一模式不断进化并推向极致而已。

2017.01.09 克里斯托住宅庭院，瓜达拉哈拉城，巴拉干，1928

3-118 Courtyard of the Gustavo Cristo House designed by Luis Barragan, Guadalajara (9 January 2017)

Among Barragan's works from the first period, my favourite and the most inspirational one is the Cristo House in Guadalajara. This was the mayor's private home, achieved with amazing maturity when the architect was only 26 years old. It is a compact urban dwelling, isolated from the urban chaos. The house is characterised by its quiet atmosphere, space and rhythm, changing scenery, three-dimensional gardens, room-like roof terraces and the elliptical arched elements at the beginning and end of the narrative. In this roughly 100 square-meter house completed in 1929, Barragan relied on the architectural approaches based on his intuitive experience to create a contemporary and wonderful house-garden/home-land, in which the living space and spirit space are integrated and interacted. Over the next 60 years, he would transform and push this model to the extreme.

3-119 路易斯·巴拉干，克里斯托住宅室内沙龙空间
（2017.01.09）

3-119 Indoor Salon Space, Gustavo Cristo House designed by Luis
Barragan (9 January 2017)

2017.01.09. 宪主斯坦巴伦宅!沙龙, 巴塞�

3-120 路易斯·巴拉干，墨西哥城，克里斯特博马厩与别墅（之一）（2017.01.15）

宅与园的分界——马厩"花园"的入口处

巴拉干一生一以贯之的"生活空间＋精神空间"。奥特加宅，巴拉干自宅，洛佩兹宅的花园是利用原有的"自然之地"（岩石、高差、林木等）并加以强化营造为"自然式花园"；克里斯特博马厩宅的花园则是在无可利用的"空白之地"生生营造的"抽象式花园"。以更抽象化的手法，用色彩、墙体，还有水面、植物进行组合，包括饮马等生活场景也进入到画面中，共同构成花园的空间。

3-120 San Cristobal Stables and the Folke Egerstrom House designed by Luis Barragan, Mexico City (1) (15 January 2017)

The boundary between the house and the garden: the entrance to the "garden" of the stables.

This represents Barragan's lifelong pursuit of the juxtaposition of "life space and spiritual space". In the gardens of the Ortega House, the Luis Barragan House and studio and the Lopez House, Barragan used and transformed the existing land (rocks, height differences, trees, etc.) to create natural gardens. However, the garden of the San Cristobal Stables and the Folke Egerstrom House is an "abstract garden" created in a "blank land". In a more abstract way, Barragan combined colour, walls, water, plants and life scenes of horses drinking with the garden space.

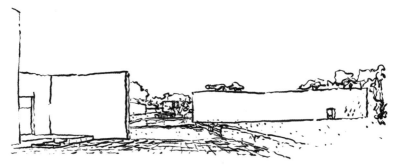

2017.01.15. 吉卫斯特博物院与刻望. 孔祥宇. 1567

3-121 路易斯·巴拉干，墨西哥城，克里斯特博马厩与别墅（之二）（2017.01.15）

由别墅起居室大窗外望。马厩，一个如现代抽象绘画般的花园。

3-121 San Cristobal Stables and the Folke Egerstrom House designed by Luis Barragan, Mexico City (2) (15 January 2017)

Looking out from the large window of the villa's living room. The stable's garden is like a modern abstract painting.

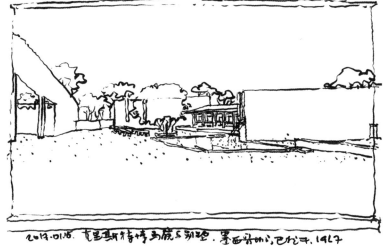

2019.01.15. 克里斯蒂博马厩与别墅. 墨西哥城, 巴拉干, 1967

3-122 路易斯·巴拉干，墨西哥城，克里斯特博马厩与别墅（之三）（2017.01.15）

水池，喷泉，厩房，彩色墙体，大门，大树，马槽／拴桩，小狗，骏马，马童。

马厩别墅尽管是 1966 年左右巴拉干年近 70 岁时的晚期作品，但在思想上和他在 26 岁时的克里斯托住宅一脉相承。而且他这种不完全采用自然元素，而是使用建筑元素，或者说通过建筑元素和自然元素的组合来营造花园的方式，实际上更适应墨西哥严峻的自然条件——巴拉干正是在这样一种土地干涸、自然元素匮乏的条件下，以建筑师的"建筑"方式和手段营造出他心中的理想"家园"。

3-122 San Cristobal Stables and the Folke Egerstrom House designed by Luis Barragan, Mexico City (3) (15 January 2017)

Pools, fountains, stables, coloured walls, gates, trees, mangers/ bolts, puppies, horses and a horse boy.

Although the San Cristobal Stables and the Folke Egerstrom House is a late work of Barragan, completed in 1966 when he was nearly 70 years old, conceptually they follow one continuous line with his Cristo House completed at the age of 26. His gardens, created by using architectural rather than completely natural elements, or a combination of architectural and natural elements, are suited to the harsh natural conditions of Mexico. Under the conditions of dry land and lack of natural elements, the architect used "architectural" means to build an ideal "homeland" in his heart.

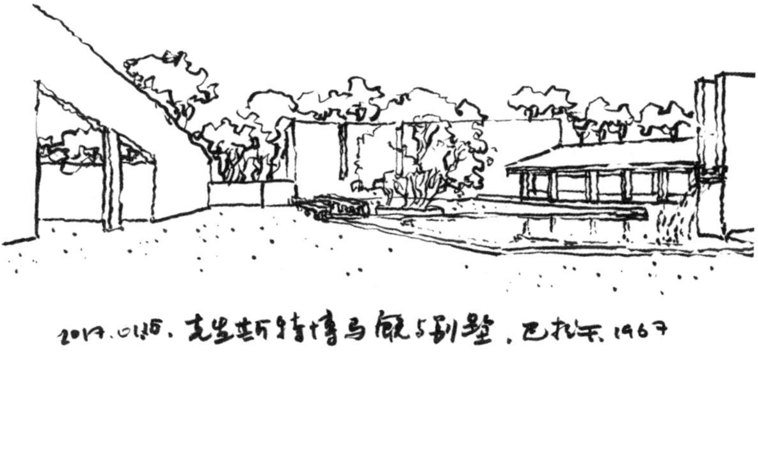

2017.01.05，克里斯特博马厩与别墅，巴拉干，1967

3-123 路易斯·巴拉干，墨西哥城，巴拉甘自宅与工作室

（2017.01.12）

　　我感觉，作为一个虔诚的天主教徒，巴拉干毕生所追求的"宁静"，在自宅中达到了前所未有的高度，但却是混合着宗教性的，渗透进了他的日常生活中、他的骨子里，是非常个人化的，甚至因此自宅中的花园都显得没那么重要了。而作为建筑师，更多时候他是要为别人服务的，正因此，他还要再往前走。

3-123 Luis Barragan House and Studio designed by Luis Barragan, Mexico City (12 January 2017)

　　I feel that the serenity sought by Barragan, a devout Catholic, reaches an unprecedented height in his home design. However, it is mixed with religion and penetrates his daily life and inner heart. It is so personal that the house's garden seems not to be important. As an architect, he must serve others more often. Therefore, he needs to go further.

2012.01.12. 巴花甘脆呢区作室. 1947

3-124 特奥蒂瓦坎，阿克托潘修道院平面图 （2017.01.12）

巴拉干说："我曾经怀着崇敬的心情寻访那些已经荒废了的修道院……我总是被那些空荡荡的回廊和孤独的庭院中的宁静深深感动。"的确，在阿克托潘修道院中，我强烈地联想起了巴拉干自宅，并不仅仅因为那高塔、开窗、体量、色彩，甚至厨房里那通向神秘之处的陡峭楼梯，而是整个布局和强烈的宗教氛围：自宅的起居室就是修道院的主教堂，卧室就是修士的居室，花园就是站在修道院拱廊平台上看到的后花园，而屋顶平台——朝向天空的花园，就象征着基督徒向往的天堂乐园。

3-124 Plan of the Actopan Monastery, Teotihuacan (12 January 2017)

"I used to look for those abandoned monasteries with reverence... I am always deeply touched by the serenity of the empty corridor and the lonely courtyard", said Barragan. Indeed, at the Actopan Monastery, I strongly think of Barragan's own house, not just because of the tower, window, volume, colour or even the steep stairs leading to the mystery in the kitchen, but the whole layout and the strong religious atmosphere. His house's living room is the main church of the monastery, the bedroom is the monks' room and the garden is the back garden seen on the arcade platform, while the roof terrace - a garden facing the sky, symbolises the paradise that christians long for.

0 4 8 12 16 20 24 28 MTS

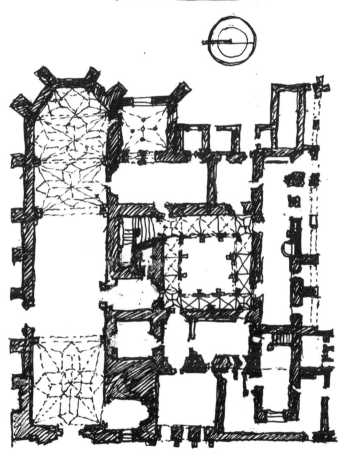

- Actopan 修道院 平面图.
  2017. 1. 12. 墨西哥之旅 巴 于

3-125 特奥蒂瓦坎，阿克托潘修道院，厨房（2017.01.11）

3-125 Kitchen, Actopan Monastery, Teotihuacan (11 January 2017)

2017.01.11. Actopan修道院 Teotihuacan

3-126 路易斯·巴拉干，墨西哥城，嘉布遣会小礼拜堂（特拉潘教堂），主堂（2017.01.13）

对比众多对应于世俗日常生活的住宅作品，嘉布遣会小礼拜堂是纯粹的"精神空间"。

3-126 Main church, Chapel for the Capuchinas (Chapel in Tlalpan) designed by Luis Barragan, Mexico City (13 January 2017)

Compared with the numerous residential works corresponding to the secular daily life, the chapel for the Capuchin as is pure "spiritual space".

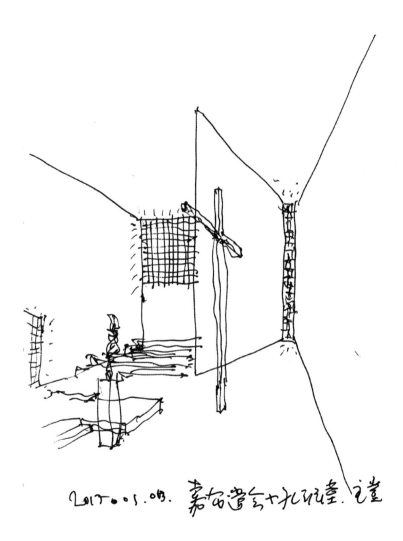

2017.05.03. 嘉定聖会十九班達·记堂

3-127 路易斯·巴拉干,嘉布遣会小礼拜堂,耳堂(2017.01.13)

3-127 Transept, Chapel for the Capuchinas designed by Luis Barragan, Transept, Chapel for the Capuchinas (13 January 2017)

2017.01.11. 嘉谷遗言扎脱境 耳臺

3-128 路易斯·巴拉干，嘉布遣会小礼拜堂，休息室
（2017.01.13）

3-128 Lounge, Chapel for the Capuchinas designed by Luis Barragan
(13 January 2017)

2017.01.13. 秦古遗今みし記堂休現堂

3-129 路易斯·巴拉干，墨西哥城，拉斯·阿博勒达斯住区景观（2017.01.15）

阿博勒达斯住区景观也是纯粹的"精神空间"，是巴拉干带给人们的"公共的私密性"。

视觉：一点透视的变异，随时间变化的 15 米高的影墙；听觉：溢水的声音；触觉：粗糙的白墙；嗅觉：桉树的清香……感受到巴拉干所追求的"平静，沉默，私密与惊愕"，以及"诗意想象力的崇高场景"。

无边界的天空，20 米的桉树，15 米的影墙，4 米的篱笆，3 米的红墙，1.8 米的落水，0.6 米的看台，0.5 米的水槽，共同将无边界的自然缩小到了人的身体尺度。

3-129 Las Arboledas designed by Luis Barragan, Mexico City (15 January 2017)

The landscape of Las Arboledas as pure spiritual space represents the "public privacy" that Barragan brought to people.

Vision: a variation of the one-point perspective, a 15-metre high shadow wall that changes over time. Hearing: the sound of overflowing water. Touch: a rough white wall. Smell: the fragrance of eucalyptus... Feeling the calmness, silence, privacy, amazement and the sublime act of poetic imagination that Barragan pursued.

The borderless sky, 20-meter high eucalyptus, 15-meter high shadow wall, 4-meter high fence, 3-meter high red wall, 1.8-meter high falling water, 0.6-meter high grandstand and 0.5-meter high sink, together reduce borderless nature to a bodily scale.

师童勤达芙手
2017.01.15. Las Arboledas 代尬景观. 巴拉干 1957

3-130 路易斯·巴拉干，墨西哥城，洛·克拉伯斯居住区，情人泉 (2017.01.15)

关于"阅读巴拉干"：在这个思绪飞扬和忙碌艰辛的年代，我们的"大匠阅读——经典建筑师思想和作品阅读系列研究"之一，精选了十二个案例作品进行了纵向分析。这些作品跨越了巴拉干的整个创作生涯；以重绘的图纸和模型作为研究的基础资料；这是一个设计研究，希望理解作品表象背后的思想内涵；我们的阅读又是带有一点"先入之见"的，是"胜景几何"的延伸，是"静谧与喧嚣"的具体化。希望每个人都能读出自己心中的那个巴拉干，"看我所见"。

3-130 Fountain of Lovers, Los Clubes designed by Luis Barragan, Mexico City (15 January 2017)

In the era full of flying thoughts, busyness and hardness, Reading Barragan is one of our research projects on classic architects' thoughts and works. This project carefully selects twelve projects (spanning Barragan's entire design career) for critical longitudinal analysis. We use re-drawn drawings and models as our basic research materials. The project is a design study, trying to understand the ideological connotation behind the appearance of these works. Our reading has preconceived ideas. It is an extension of "integrated geometry and poetic scenery", an embodiment of "tranquillity and noise". I hope that everyone can read Barragan from their heart. "See what I saw".

R017pl.15 · 情人泉, LOS Clubes住区 墨西哥城 巴拉干1958

3-131 特奥蒂瓦坎，由太阳神庙看月亮神庙（2017.01.11）

　　坐在"太阳"上看"月亮"。极端体现人工独立感的一座座石砌金字塔形玛雅神庙，坐落于连绵的山脉与原野之中。

3-131 Looking the Temple of the Moon from the Temple of the Sun, Teotihuacan (11 January 2017)

　　Sitting on the Temple of the Sun to see the Temple of the Moon. A number of stone-built pyramid-shaped Mayan temples with an extremely artificial sense of independence locate in the rolling mountains and wilderness.

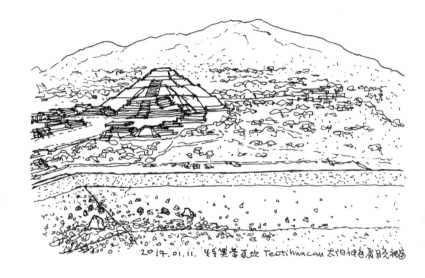

2017.01.11. 在墨番哥城 Teotihuacan 太阳神庙看日光相影

3-132, 133, 134

勒·柯布西耶，日内瓦湖，母亲住宅（2017.05.08）

　　由入口庭院至起居室内横长窗对日内瓦湖山的框景、屋顶平台到猫的"观景"小平台与狗的墙边穴居，特别是临湖高起一段石造围墙，开窗洞，洞边一桌四椅，大树下坐而观望湖山胜景，是整体空间设计不可或缺的构成部分，亦可将被建筑周边的围墙包裹的庭院视为建筑主体之外没有屋顶的内向空间。柯布西耶在这个小建筑中的设计与中国园林的空间营造方式非常相近，使得庭院、湖山与建筑相互因借为一体。

3-132, 133, 134 Villa Le Lac designed by Le Corbusier, Lake Geneva (8 May 2017)

　　The entrance courtyard, the horizontal window in the living room framing Lake Geneva and mountains, the roof platform, the small platform for the cat to view, the dog's burrow against the wall, and especially the stone wall rising over the lake, with the opening window-hole whose side is a table and four chairs, offering an opportunity to sit under a tree and view the lake and mountains, these are indispensable parts of the overall space design. The courtyard, surrounding the bounding walls, can also be regarded as an inward space without a roof outside the main body. Le Corbusier's design of this small building is very similar to the space created in Chinese gardens, intertwining the courtyard, lake and mountains with the architecture.

3-135 阿尔瓦罗·西扎，波尔图，海边茶室（2019.05.25）

　　西扎最让人望尘莫及的是，不到 30 岁已有成熟大作，年过 80 仍然活力丰沛；而在长达 50 多年的职业生涯中，西扎始终保持稳定少变的设计语言，却不断给人惊喜。

3-135 Waterside Tea House designed by Alvaro Siza, Porto (25 May 2019)

　　The most fascinating thing about Siza is that he had mature great work at the age of 30, and still had a lot of energy into his 80s. In the course of more than 50 years of his professional career, his design language is always stable and less changeable. However, it has continued to surprise people.

3-136 阿尔瓦罗·西扎，杭州，中国美术学院，中国国际设计博物馆（2018.12.29）

园林般的"多点透视"空间。

3-136 China International Design Museum designed by Alvaro Siza, China Academy of Art, Hangzhou (29 December 2018)

A garden-like space with multi-point perspectives.

2018.12.29. 杭州 中国美院包豪斯家兴博物馆
A. 西扎

• 围材搭叠的透视层面"空间

3-137 彼得·卒姆托，科隆，Kolumaba Kunst 博物馆

（2015.10.28）

遗址如自然的"花园"，柱林、游廊、透光的砖墙如园墙上巨大的漏窗。

3-137 Kolumaba Kunst Museum designed by Peter Zumthor, Cologne (28 October 2015)

Ruins here looks like a natural garden. Pillars, a corridor, light-transparent brick walls look like huge leaking windows on the garden wall.

2015.10.28. KOLUMBA MUSEUM
KUNST-
Kolumbastraße 4. D-50667 Cologne (Peter Zumthor)
(Köln)   1997~2007

3-138 关颂声、杨廷宝，南京中山陵，音乐台（2015.07.22）

可称是世界上最省钱的"音乐厅"，但据说音乐欣赏效果极佳。人造扇形台地草坪地形，辅以环廊、水池、舞台及反射音壁，周围山林包裹，犹如世外降临的天国乐（yuè）园，景观建筑与音乐功能结合完美的大作。

3-138 Music Station designed by Guan Songsheng and Yang Tingbao, Nanjing Sun Yat-sen Mausoleum (22 July 2015)

It may be the most cost-effective "concert hall" in the world, but it is said that it produces an excellent music appreciation experience. The man-made scalloped and terraced lawn, complemented by the circular corridor, pool, stage and reflective sound wall, is surrounded by mountains and forests. It looks just like a music garden beyond this world, where the combination of landscape architecture and music function produces a perfect masterpiece.

南京中山陵音乐台. 美纪事 杨定宝 (1983) 2015.07.22

3-139 董豫赣，江西，耳里庭客舍 （2019.06.02）

地形、地景与场景。董老的绵密营造——几段猪圈残基引发的神奇：凸凹、藏露、高下、曲折，疏可走马、密不透风，体宜因借、居景互成。处处匠心，就服董老。

3-139 Courtyard within Earshot designed by Dong Yugan, , Jiangxi Province (2 June 2019)

Terrain, landscape and scenes. Dong's meticulous design and construction build a magical place, composed of several segments of a pigsty. It is convex and concave, hidden and visible, high and low, with twists and turns. It is appropriately both loose and tight, borrowing scenery that interacts with the villa. Ingenuity is everywhere. I admire old Dong.

2019.06.02. 耳里庭舍宅/ 葛建程

3-140 科隆，大教堂（2015.10.31）

与人尺度对应的构件组装成山一般升腾而起的建筑体量。黑色，沉默。

3-140 Cologne Cathedral (31 October 2015)

The components, corresponding to the human scale, are assembled into a building that rises like a mountain. The image is black and silent.

585

2015.10.31. KÖLN DOM
科隆大教堂

3-141 芝加哥，Paloma 公寓外望（2016.05.11）

芝加哥七日，重新发现密斯的魅力：less 确实 is more，沉静之景 in city scape。……话说密斯当年在芝加哥这是接了多少大活儿啊，而且基本都是标准做法，性价比倍儿高。西尔斯虽不是密斯的，却是集束密斯式的。

3-141 Looking out from the Paloma Apartments, Chicago (11 May 2016)

Seven days in Chicago, rediscovering the charm of Mies: less is indeed more, a quiet scene in the city scape... I wonder how many big projects Mies has completed in Chicago. These cost-effective commissions were basically finished by taking a standard approach. Although the Sears Tower was not designed by Mies, it is a cluster of Mies-style buildings.

2016.05.11 삿뽀로 paloma 누힐에서 望

3-142 上海，由环球金融中心顶层酒廊眺望，夕阳中的黄浦江两岸城市建筑群，近景为上海中心、金茂大厦和东方明珠广播电视塔（2019.03.26）

都市胜景。

3-142 Overlooking the urban buildings on both sides of the Huangpu River at sunset from the top floor lounge of the World Financial Center, Shanghai. Nearby buildings include the Shanghai Centre, Jinmao Tower and the Oriental Pearl TV Tower (26 March 2019)

Urban poetic scenery.

上海/2019.03 都市阳台

3-143 北京，由工作室平台东望，雨后晴朗天空下的城市，可见白塔寺、故宫、国贸、中国尊等（2019.05.26）

都市胜景。

从工作了十年的工作室搬到这里，想到：时光和生活可以赋予空间一种特别的东西，我称之为特殊的"自然"，让我们珍视。

3-143 Looking east from the platform of our studio, the city under the clear sky after the rain, with views of the Baita Temple, the Forbidden City, Beijing World Trade Tower and the CITIC Tower (26 May 2019)

Urban poetic scenery.

Moving here from a studio where we worked for ten years, I think that time and life can give space a special thing. I call it a special "nature". Let us cherish it.

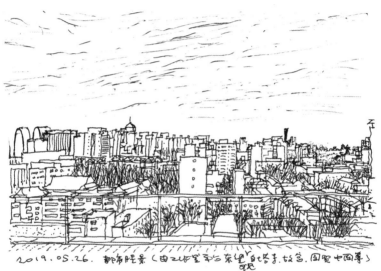

2019.05.26. 都市暗景（由乙住室窗台东望，自塔于、故宫、国宾七面丰）

居 / 造（日常、研究、场地与设计）

　　由"游 / 憩"，到"行 / 旅"，再到"望一栖"，获得的所有感悟延伸至"居 / 造"的日常之中，这里既有日常生活中的观察与思考，亦有与《大匠》、"园林"等设计性研究相关的记录与感悟。既有前往项目场地踏勘时的即时触动与发现，亦有设计探索过程中的灵感捕获与呈现，乃至细节研究与表达。想与做交织，研究与设计同行，边思考边实践——希望自己能成为当代的、现实中的、日常的、为人的理想生活空间营造者。针对着具体项目的场地和营造，草图成为真正的草图，"图语"成为实践行动旅程的构思和落地记录。由南京"瞬时桃花源"作为当代现实中的"理想实验"，到涵盖不同主题及共同策略的当代现实中的"理想实践"，"现实理想空间营造范式"提示出一种可能连接悠久传统、当代现实与不可见未来的建筑学图景。"胜景几何"——风水形势、人作天工、结构场域、叙事空间、情境胜景。营造与自然交互的建筑，建筑成为人与自然交互的中介——"自然"的建筑。坚固，适用，自然，愉悦。

Living
(Daily, research, site and design)

From wandering/recreation to walking/travelling and viewing/ inhabiting, all of the sentiments gained extend to the daily life of living/building. There are observations and thoughts in daily life, as well as records and sentiments related to design studies such as *Reading on Masters* and *Gardens*. There are both immediate revelations and discoveries during the exploration of project site, in addition to the capture and presentation of inspiration during the design exploration process, as well as research and expression of details. With the interweaving of thinking and making, the juxtaposition of research and design and the parallel between meditating and practising, I hope to become a contemporary, realistic, daily builder for peoples' ideal living spaces. For each specific project site and construction, the sketch becomes a plan and the graphic languages become the concept and record of the practical action and journey. From the "ideal experiment" in the Nanjing "Instantaneous Peach Garden", to the "ideal practice" in contemporary reality covering different themes and common strategies, the "paradigm of building realistic ideal space" that I have summarised suggests a possible architectural picture that connects tradition, contemporary reality and the invisible future. Integrated Geometry and Poetic Scenery: Feng shui situation, the man-made and nature-work, the structural field, the narrative space, the situation and the poetic scenery. To create a building that interacts with nature means that architecture becomes the intermediary of human-natural interaction - the "natural" architecture, which is firm, practical, natural, and enjoyable.

4-001 一只特立独行的鸡（2017.01.27）

4-001 A maverick cock (27 January 2017)

2017. 01. 27

4-002 三叹随感（2019.06.02）

　　每读中国文献，总有三叹，叹自己未尝具备三种能力：一曰诗词，一曰文言，一曰繁体。身为此文化中人却半遮半挡于彼文化信息之外，一知半解，遑论述作，觉若幼齿儿童，殊感汗颜，汗颜汗颜。中国语言文字乃至艺术及于建筑，均各完备自立系统，又潜潜合于一体思想哲学，在在可称"文明"二字，实应宝贵而不弃者也。现代诸项改良革命，如白话标点句读，方便实用自不待言，而失却音韵节奏、清简爽利、古雅神秘之质感——一句话，不高级了。身为现代人自然不可复古，但鱼与熊掌兼而得之，不好吗？

4-002 Three sighs and personal comments (2 June 2019)

　　Every time I read Chinese literature, I emit three sighs, for my inability to master poems, classical Chinese and traditional Chinese characters. As a Chinese, I am almost separated from this culture, having a superficial knowledge of it, let alone the ability to write works. I am like a young child, feeling awkward.

　　The Chinese language, characters and even art and architecture are all self-sustaining independent systems, integrated into one ideological philosophy. Worthy of being called a civilisation, they should be viewed as precious and not be abandoned. Modern reforms and revolutions, such as vernacular Chinese and punctuation marks, are convenient and practical, but they are losing their rhythm, simplicity and quaint mysterious nature. In a word, they are not advanced any more. As modern persons, we cannot go back to the past, but isn't it good to have "both fish and bear's paw at the same time"?

歸止爰閤去門立門啟闔先藏
大藏
孔城圓通父攜陷子死葬

2019.06.02
篆顏真卿《爭座文稿》

4-003 摹写 [明] 文徵明《拙政园三十一景图》之"听松风处"（2019.04.30）

沉浸式空间。

《园冶·兴造论》开篇说"妙在得体合宜"，后面又讲"巧于因借，精在体宜"，计成用"精""妙"形容"得体"，是很高的评价。人到中年，越来越觉得"得体"的重要。

在北京服装学院听鲁安东教授讲座《从拙政园到留园：沉浸式空间的兴起与衰落》并参与讲后讨论，边思考边发言中，我对"因借体宜"又有了新的理解："因借"可能并非是指借用外物或外景，而是指的凭借某种"中介物"，如建筑（亭）、景物（梧竹）和文字（诗题）；"体宜"也并非是指得体合宜，而是作为体验或空间使用主体的人（不同时代的相似身体）获得身心一体的沉浸式感受（如"梧竹幽居"）。即通过"因借"/凭借中介物而达到主体身心沉浸的"体宜"。

4-003 Imitating pavilion for listening to the wind of pine (*Ting songfeng chu*), one of the *Thirty-one Scenic Spots of the Humble Administrator's Garden* drawn by Ming Dynasty painter Wen Zhengming (30 April 2019)

Immersive space.

At the beginning of *Yuanye Xingzaolun*, the author Ji Cheng wrote that "ingeniousness lies in appropriateness". He later said that "borrowing scenery" is clever and appropriateness is refined, using "refine" (*jing*) and "ingenious" (*miao*) to describe "appropriateness" (*deti*). This is a very high evaluation. Reaching middle age, I become more aware of the importance of appropriateness. While listening to Lu Andong's lecture *From the Humble Administrator's Garden to the Liu Garden: the Rise and Decline of Immersive Space* at the Beijing Institute of Fashion Technology and participating in the discussion, I gained a new understanding of *yin jie ti yi*. *Yin jie* may not mean borrowing outside objects or scenery, but referring to certain intermediaries, such as architecture (pavilion), scenery (bamboo) and text (poem). *Ti yi* does not mean appropriateness, but the subject of experience or the space user (a similar body at different periods) who gains an immersive feeling of body-mind integrity (such as in the Wuzhuyouju Pavilion). Through some intermediary elements (borrowing), the subject gains physically and mentally immersed appropriateness.

2019.4.30《拙政园三十一景图》09木与风处
[明]文徵明 "沉浸式空间"

4-004 摹写 [明] 文徵明《拙政园三十一景图》之 "槐幄"

（2019.04.30）

沉浸式空间。

安东教授将自己的直觉抽丝剥茧变成精确的理论性判断，讲座中 "沉浸式空间的衰落"，与陶特、格氏、柯式对桂离宫乃至日本园林的现代主义 "误读"，以及日本洄游式园林所带有的现代主义美学有某种相似之处，也促使我思考以下问题：这一话题对建筑学的当代意义和现实意义？人作为空间使用的主体，那么对应的是当代什么样的人——是否更多是指使用公共空间的人？以及当代什么样的私密性——是否是一种 "公共的私密性"（如在佛光寺的空间体验）？什么将替代文字作为 "主体诱导" ——文人 - 文字是否被大众或者个体 - 数字媒体取代？园林与寺庙、宫殿、民居的一体性意义：是否不过是对应 "沉浸式空间" 中不同层次的主体？

4-004 Imitating Big locust tree (*Huai wo*), one of the *Thirty-one Scenic Spots of the Humble Administrator's Garden* drawn by Ming Dynasty Painter Wen Zhengming (30 April 2019)

Immersive space.

Through detailed analysis, Andong's own intuition has become a precise theoretical judgment. The decline of immersive space discussed in the lecture, the misunderstanding of the Katsura Imperial Villa and the Japanese garden by Bruno Taut, Walter Gropius and Le Corbusier from the point of view of modernism, and the modernist aesthetics embodied in Japanese gardens share some similarities, prompted me to think about the following questions. What is the contemporary meaning and practical significance of these thinking and discussion on architecture? The main body of using space is people, then what kind of people in contemporary era? Does this refer to people who use public space? What kind of privacy is it: public privacy (such as the spatial experience at the Foguang Temple)? What is the substitution of text as a "subjective induction"? Will literati-text be replaced by the mass or individual-digital media? The integral meaning of gardens, temples, palaces and dwellings: are they just corresponding to the subjects at different levels of "immersive space"?

2019.04.30 《拙政园三十一景图》之"槐雨亭"
[明] 文徵明 "沉浸式空间"

4-005 遥想江南四月天 （2016.04.03）

清明。雨过尚未天青，风含三江新绿。

4-005 Recalling April's Jiangnan (3 April 2016)

　　Qing Ming Festival. The sky has not yet cleared after the rain, and the wind contains a new green sense.

雨过高来　天青风含三江新绿

二O一六.1郝明

4-006 蔡国强，墨尔本 NGV 艺术大展："鸟云"——瞬间的山水与永恒的卫士（2019.06.03）

态度、观念、形式。蔡国强问：（社会政治等之外，）"艺术到底怎么样？"发人深省。语言与意境，道与器。其实语言才是前提。意境固然重要，语言确实如此重要；语言之中，形式如此关键，意味深长、深思熟虑的形式。

4-006 Clouds of Birds, *The Transient Landscape* and *Guardians of Immortality* Cai Guoqiang, Melbourne NGV Art Exhibitions (3 June 2019)

Attitude, concept, form. Cai asked what art is indeed, aside from its social and political functions. This is thought-provoking. Considering language and artistic conception, Dao and utensils, language is the premise. Although artistic concept is important, language is also important. Form is very crucial to language, particularly meaningful and deliberate form.

607

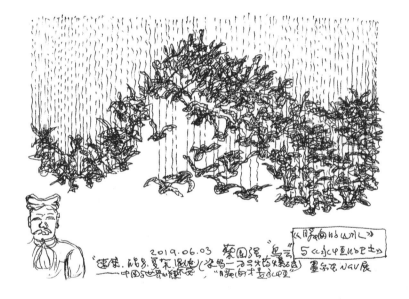

2019.06.03 蔡国强 "画云"

《月落的山川》

5《永恒的迷土》

墨尔本 NGV展

4-007 阿尔瓦罗·西扎，意大利加埃塔市立当代艺术美术馆（拉齐奥）"Personaggi"个展，《梅萨》（2019.05.22）

西扎爷的独门"观法"：Juan Navarro Baldeweg 发表于《Domus》国际中文版 085 期的文章《雕塑家的凝视》评介了 2014 年阿尔瓦罗·西扎在意大利拉齐奥加埃塔"Personaggi"个展中从未被展示过的雕塑作品。文章作者认为，对于绘图和雕塑的不懈追求，是西扎建筑的基础和灵感起源，并且西扎的草图性绘画具有一种独特的动态特性和不同观察点的共存，这犹如雕塑家不断移动他凝视作品的目光，这种使用手眼协作与感知之间的循环往复，是一种"最简单工具"的工艺，使得心中的事物"可见从而定义现实"。

4-007 *Mesa*, Alvaro Siza, solo exhibition of *Personaggi* at the Museum of Contemporary Art (Lazio), Gaeta, Italy (22 May 2019)

Siza's unique method of viewing. Juan Navarro Baldeweg published an article in *Domus* (Chinese edition, no. 85) called "The Gaze of the Sculptor". It reviewed Alvaro Siza's sculptures (never displayed before) shown at the 2014 solo exhibition, *Personaggi*, in Lazio, Italy. The author of the article explained that Siza's relentless pursuit of drawing and sculpture was the basis and inspiration of his architecture. Siza's sketchy paintings have a unique dynamic character, and the coexistence of different observation points, just as if the sculptor is gazing at his/her work while moving his/her sight. The reciprocating process between hand-eye collaboration and perception is the craft of the simplest tool that makes things in the heart visible, and thus defines reality.

《梅薩》

2019.05.22. Alvaro Siza 因佳裡家的凝视

4-008 阿尔瓦罗·西扎，意大利加埃塔市立当代艺术美术馆 "Personaggi" 个展，《尼诺》和《皮诺》（2019.05.22）

我想，这无疑是解读西扎建筑作品中那令人着迷的雕塑性、暧昧感、漫游空间乃至场所特征的最重要密码。正如巴西建筑大师保罗·门德斯·达·洛查在《众生之城》中所言："建筑永远是关于我们想要的和我们已经做到的两者之间的对话……在它的诗意表达中，建筑转化为一种描绘将要出现的事物的语言……对欲望的理解隐藏在形式中……在建筑中存在着介于理念与实物之间的生命力……它的目的是将理性与自由的关系具体化。"

4-008 *Nino* and *Pino*, Alvaro Siza, solo exhibition of *Personaggi* at the Museum of Contemporary Art, Gaeta, Italy（22 May 2019）

I think this is undoubtedly the most important code to interpret the fascinating sculptural form, sensation and roaming space and even the place characteristics in Siza's architectural works. As the Brazilian architect Paulo Mendes da Rocha said in *The City of All Lives*, "Architecture is always a dialogue between what we want and what we have done... In its poetic expression, architecture is transformed into a language that describes the things that will appear... The understanding of desire is hidden in form... architecture maintains the vitality between ideas and objects... Its purpose is to materialise the relationship between reason and freedom."

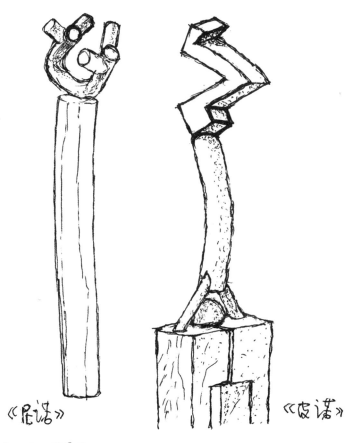

《皮诺》

《皮诺》

2019.05.22
Alvaro Siza. 西扎. 托雷果加寅塔 "Persovaggi" 个展

4-009 阿尔瓦罗·西扎，里斯本，Manuel Cargaleiro 基金会
（2019.05.22）

顺着西扎雕塑家般凝视的目光，我们能够感知到半个多世纪以来始终持续在他众多建筑作品中的独特诗意；然而对比之下，很多振振有词、逻辑精确的当代建筑作品却无法令人产生由衷的感动。这不禁使我反省：在当代人们生存环境的普遍危机和困惑中，在不断更新的各种技术、观念、理论和适用、坚固、可持续等之外，建筑将如何为人们营造出与时空体验紧密关联的愉悦和诗意？这是当代建筑师们仍然亟需思考和解答的本原性命题。

4-009 Manuel Cargaleiro Foundation designed by Alvaro Siza, Lisbon (22 May 2019)

Following the gaze of Siza similar to that of a sculptor, we can perceive the unique poetics that have been continually demonstrated in many of his architectural works for more than half a century. However, many of the most plausible and logically accurate contemporary architectural works do not present a heartfelt impression. I cannot help but reflect on myself: amid the omnipresent crisis of the contemporary human living environment and confusion, beyond the constantly updated technologies, concepts and theories and their applicability, firmness and sustainability, how can architecture create pleasure and poetics for people that are closely related to the time and space experience? This is the ontological proposition that contemporary architects still need to think about and answer for.

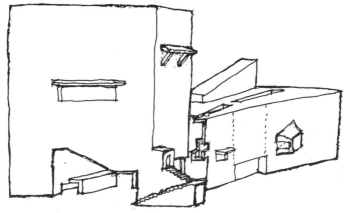

2019. 05. 22.
Alvaro Siza, Manuel Cargaleiro Foundation, Lisbon

4-010 仿写关良《贵妃醉酒》（2019.03.08）

宾夕法尼亚大学沃顿中国中心，关于"技术、文化、未来"的建筑师对话。我提了两个问题：1. 什么技术及其发展是为建筑学所真正需要的？ 2. 当我们拥有和使用某种被（高度）发展的技术时，我们会（同时）失去什么？

4-010 Imitating Guan Liang's *The Drunken Beauty* (8 March 2019)

At the architects' dialogue on "technology, culture and the future" at the University of Pennsylvania's Wharton China Center, I asked two questions: (1) What kind of technology or its development is really needed by architecture? (2) What do we lose when we own and use a kind of developed technology?

贵妃醉酒

4-011 南京，中国建筑新人赛，"逢竹记"——从茶到室

（2018.08.19）

东南大学中国建筑新人赛评图，我最喜欢的《逢竹记》未能入围。一点感想：我们的建筑教育（乃至职业环境）中对概念、空间、形式等的过度重视而对结构的忽略和轻视；对建筑本体的过度强调而对生活感知、意象表达的抑制和弱化；还有对理性、逻辑的过度鼓励而对感性、偶然的冷落和批评。

4-011 Meeting Bamboo (*Fengzhu Ji*) — *From Tea to Room*, China Architecture Newcomers Competition, Nanjing (19 August 2018)

Reviewing the China Architecture Newcomers Competition at Southeast University, my favourite project, *Fengzhu Ji*, failed to be shortlisted. A little thought: our architectural education (and even the professional environment) pays too much attention to concept, space, form, etc., while neglecting and scorning structure. It over emphasises the architectural ontology but suppresses and weakens the expression of life perception and image. There is also excessive encouragement to use reason and logic, while deserting and criticising sensibility and contingency.

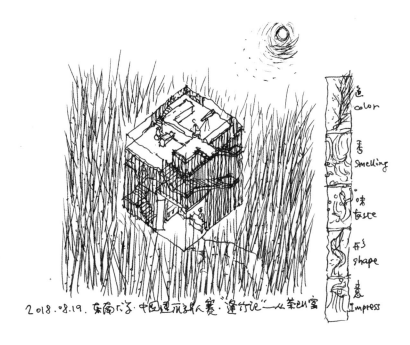

2018.08.19. 东南大学·中国建筑学会大赛·"建行记"——从茶到宝

色 color
香 smelling
味 taste
形 shape
意 impress

4-012 南京，花露岗，"瞬时桃花源"场地（之一）(2015.05.23)
废基，麦田，孤树，城墙，新城。

4-012 Site of the Instantaneous Peach Garden, Hualugang, Nanjing
(1) (23 May 2015)

Waste base, wheat field, lone tree, city wall, new town.

2015-05-23
南京 花露岗

4-013 南京，花露岗，"瞬时桃花源"场地（之二）（2015.05.23）
麦田，城墙，孤树，树丛，荒废的老房子。

4-013 Site of the Instantaneous Peach Garden, Hualugang, Nanjing (2) (23 May 2015)

Wheat field, city wall, lone tree, bushes, abandoned old houses.

2015-05-23 南京 花露岗

4-014 南京，花露岗，"瞬时桃花源"计划草图 (2015.06.25)

格物：格一下物，致一点知。阁、亭、廊、塔与废基、孤树、城墙、"山石"。

4-014 Sketch of the plan of the Instantaneous Peach Garden, Hualugang, Nanjing (25 June 2015)

Investigation and research of things (*gewu*): exploring a few things, gaining a little knowledge. I see the pavilion, corridor, pagoda and waste base, lone tree, city wall and "mountain stone".

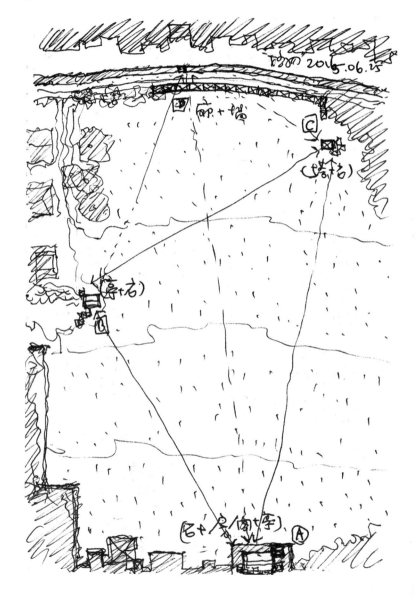

4-015 南京，"瞬时桃花源"，远望台阁（2015.07.10）

4-015 Overlooking the Terrace Pavilion, the Instantaneous Peach Garden, Nanjing (10 July 2015)

暖暖时和台客厅

2015·7·10 台南·左镇看

4-016 南京，"瞬时桃花源"，台阁内坐而外望（2015.07.10）
格物与述说。

4-016 Looking outside from the Terrace Pavilion, the Instantaneous
Peach Garden, Nanjing (10 July 2015)
Gewu and description.

2015.07.10 S面, 花暖阁, 吃冷时水化花因

4-017 南京，"瞬时桃花源"，远望树亭（2015.07.10）

4-017 Looking at the Tree Pavilion, the Instantaneous Peach Garden, Nanjing (10 July 2015)

怀柔田地与风光写

2015.07.10 初季·花草尚

4-018 南京，"瞬时桃花源"，树亭内坐而外望（2015.07.10）
幻象与现实。

瞬时桃花源，它并不只是我的，而是属于所有为它付出帮助、劳动，在其中沉思和体验的人，属于南京城西南隅这块场地及其沧海桑田的历史。当它拆除复归场地原状，这些影像就将成为它曾经发生和存在的证明，而更为深刻的记忆则将留存在曾经到过这里的人们心中。这就是"瞬时"的意义吧：愈是短暂，愈可久长。

4-018 Looking outside while sitting inside the Tree Pavilion, the Instantaneous Peach Garden, Nanjing (10 July 2015)

Illusion and reality.

The Instantaneous Peach Garden project is not just mine, but belongs to all those who give help, labour for, contemplate and experience it. It belongs to the site of the southwest corner of Nanjing City and its long history. When it was dismantled and the site reverted to its original condition, these images became proof that it had happened and existed, and a deeper memory remains in the hearts of those who had been here. This is the meaning of instantaneous: the shorter it is, the longer it can be.

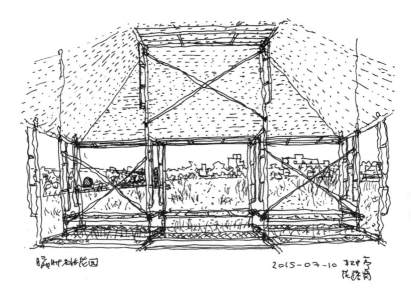

日本岬观光花园

2015-07-10 松亭
陈露角

4-019 南京，"瞬时桃花源"，树亭草图 （2015.07.11）

　　吾亦微园。蒙豫赣兄现场启示，于葛明兄所作的微园中墨绘树＋亭草图——若临池设亭，则树、池、亭与老房边界围合，而成"微园"。

4-019 Sketch of the Tree Pavilion, the Instantaneous Peach Garden, Nanjing (11 July 2015)

　　I also designed a micro garden. Inspired by Dong Yugan, I used inks to redraw a sketch of the Tree Pavilion in the micro-garden designed by Ge Ming. If the pavilion was set up near the pool, then the trees, pool, pavilion and the edge of the old house would surround to make another "micro-garden".

4-020 南京，花露岗，"瞬时桃花源"拆除计划

（2015.08.01—02）

南京花露岗"瞬时桃花源"于 2015 年 8 月 1 日被拆除，存世 24 天。物返各处，色空合一。尘归尘，土归土，整个过程碳排放总量暂计为零。再见！不带走一片云，不留下一丝痕，只有夏草在疯长。

瞬时桃花源的设计和建造手记。无论多么微小、快速、短暂的设计建造，都有一个具体而微的过程：真实的现场，与工匠合作、相互学习，劳作、攀爬、随机应变，为自己所经历和做到的一切而愉悦、感动。

4-020 Demolition plan for the Instantaneous Peach Garden, Hualugang, Nanjing (1, 2 August 2015)

The Instantaneous Peach Garden in Hualugang, Nanjing City was demolished on 1 August 2015. It had survived for 24 days. Everything goes back to its original state, and form and emptiness are combined. The dust is returned to dust, and the soil is returned to the soil. The total carbon emissions from the whole process are nearly zero. Goodbye! Do not take a cloud and leave no trace. Only summer grass is growing crazily.

The design and construction notes of the Instantaneous Peach Garden: no matter how small, fast and short-term the design and construction are, there is a specific and detailed process: the real site, cooperation with the artisans, mutual learning, labour, climbing, adapting and feeling pleasant then moving for everything that we have experienced and done.

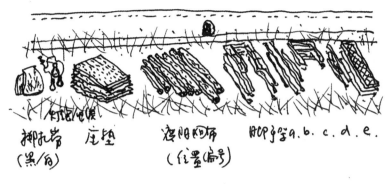

钉泡/电线

挪扎带　　座垫　　　遮阳网布　　　那叶子等 a. b. c. d. e.

（黑/白）　　　　　　　（住墨/偏灰）

· 台面 · 扎叶亭 · 墙布 · 女塔
　（台上）　（扎坏）　（墙前）　（草丛中）

　　　　拆下　　　　　　　随时拆
· 分别归类堆放. 拍照（背景环境不同拍问.）

· 座垫. 遮阳网布. 剩余挪扎带. 钉泡/电线等 打包运回
　　　　　　　　　　　　　　　　　　　　北京.

· 拆除. 摆放过程全程录像.

· 2015. 8.1～8.2

临时挑战馆墙拆除计划

台上(南望)　　扎下/塘边　　墙前　　草丛(东北望)　　台上(北望)

4-021 圣诞快乐＋新年快乐 (2015.12.24)

2015 年做了两件重要的事情，一是一个月内建造并拆除的南京"瞬时桃花源"，一是耗时数月勉力完成的《静谧与喧嚣》。

文字写作对我来说殊为不易，把我自学习建筑以来不同阶段的思考和实践进行了归纳和整合，开篇文章《静谧与喧嚣》是对"胜景几何"特别是"空间诗意"涵义的深化思考，我戏称之为"胜景几何 2.0"。这本书是"想"，瞬时桃花源是"做"，想是做的向导，做是想的实验，作为研究，需要专一从而深入；而作为实践，则需要综合从而全面解题，即一种"任何局部都不可预见的整体性"。对我来说，建筑这东西，最重要是想做合一，希望在未来的工作中，能显示出它们的意义。

4-021 Merry Christmas and Happy New Year (24 December 2015)

In 2015, two important things happened. One was the Nanjing Instantaneous Peach Garden, which was built and dismantled within a month. The other was the book that was completed within a few months, *Tranquillity and Noise*.

Writing is very difficult for me. I have summarised and integrated my thinking and practice at different stages since studying architecture. The article, entitled "Tranquillity and Noise" reflected my deeper thinking about the meaning of "Integrated Geometry and Poetic Scenery", especially the spatial poetics. This book was about thinking, whereas the Instantaneous Peach Garden project was about doing. Thinking is a guide for doing, and doing is an experiment of thinking. Research must be specific and deep, whereas practice must be integrated and solve problems comprehensively, or achieve an integrity unforeseen from any part. For me, the most important aspect of architecture is the integration of thinking and making. I hope that future work will show its meaning.

瞬時

桃花園

鳳凰台上鳳凰遊
鳳去台空江自流
吳宮花草埋幽徑
晉代衣冠成古丘
三山半落青天外
二水中分白鷺洲
總為浮雲能蔽日
長安不見使人愁

李白登金陵鳳凰
台詩一水中分
中秋世心相宜書

4-022 格物上海展，"瞬时桃花源"展览方案草图（2015.10.07）

格物：从南京到上海。"月光"下的瞬时桃花源，四只新加的"椅亭"，树亭变成半（透明）屋（2016.03.12）。何处不是花露岗？

4-022 Sketches of the exhibition plan of the Instantaneous Peach Garden, Shanghai Exhibition of *Gewu* (7 October 2015)

Gewu: From Nanjing to Shanghai. Instantaneous Peach Garden under the "moonlight". There are four newly added "Chair Pavilions", and the Tree Pavilion becomes a semi-transparent house (12 March 2016). Where is not Hualugang?

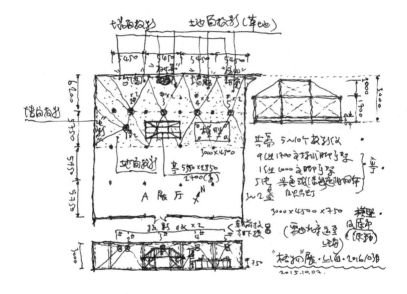

4-023 怀柔，水长城书院现场草图（之一）（2016.03.23）
由井亭望对山长城。山里桃花已盛开。

4-023 On-site sketch of the Waterside Great Wall Academy, Huairou (1) (23 March 2016)

　　Looking at the Great Wall from the Well Pavilion. The peach blossoms are already in full bloom in the mountains.

2016.03.23 怀柔水长城书院·由井亭望对山长城.

4-024 怀柔，水长城书院现场草图（之二）（2016.03.23）
由栗树林平台西眺。

4-024 On-site sketch of the Waterside Great Wall Academy, Huairou
(2) (23 March 2016)
Looking west from the chestnut forest platform.

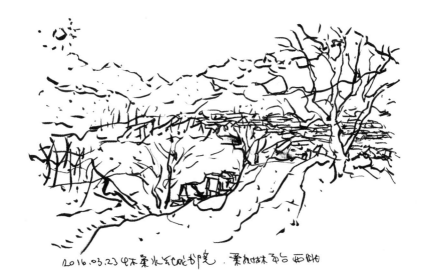

4-025 怀柔，水长城书院现场草图（之三）　（2016.03.23）
幽谷。世外桃源。

4-025 On-site sketch of the Waterside Great Wall Academy, Huairou
(3) (23 March 2016)
　　Secluded valley: a paradise.

4-026 怀柔，水长城书院现场草图（之四）（2016.03.23）
由长城残垣洞口远眺。

4-026 On-site sketch of the Waterside Great Wall Academy, Huairou (4) (23 March 2016)

Overlooking from the hole in the ruins of the Great Wall.

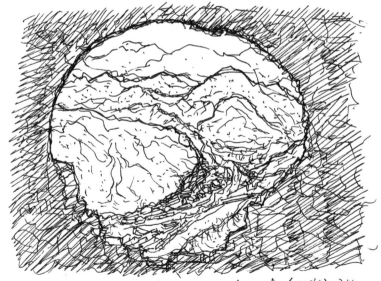

2016.03.23. 怀柔水长城书院. 由长城这眺

4-027 延庆，世界园艺博览会国际馆，竞赛方案草图
（2016.04.27）

以房为山。也想营造一个"桃花源"。

4-027 Sketch of the plan for the International Pavilion Competition,
International Horticultural Exposition, Yanqing (27 April 2016)

Taking the "house" as the "mountain", I also want to create a
"Peach Garden".

世界園艺博览会
国际馆

4-028 厦门，胡里山，厦门音乐中心场地（2015.11.07）
岸，海，岛，山。

4-028 Site of the Xiamen Music Center, Huli Hill, Xiamen (7 November 2015)

Shore, sea, island, mountain.

2013.11.07 厦门. 胡里山. 百乐山田园山.

4-029 厦门，胡里山，厦门音乐中心设计构想草图

（2015.11.07）

建筑层层跌落，构成新的海岸地形。

4-029 Design concept sketch of the Xiamen Music Center, Huli Hill, Xiamen (7 November 2015)

The building fell down level to level to form a new coastal terrain.

4-030 惠灵顿，中国驻新西兰大使馆场地观察与记录（之一），自西东望（2015.10.20）

原来小块产权用地上的房屋拆除后留下的几何状台地组合，均已被荒草植物覆盖得边界模糊起来，东面是城市、远山丘陵及城市标志之一——总督府的尖塔。

4-030 Site observation and record (1), looking east from the west, Chinese Embassy in Wellington, New Zealand (20 October 2015)

The houses built on the small lands with property rights have been demolished. The site is a combination of geometrical terraces covered by grass and plants, which blurs the site borders. In the east is the city, the hills and one of the city's symbols, the spire of the Governor's Mansion.

2015.10.20. Wellington 从信用地自西亭望

4-031 惠灵顿，中国驻新西兰大使馆场地观察与记录（之二），自南北望（2015.10.20）

由坡地南端最高处的台地北望，Rugby 街 73 号如揳入场地的"钉子"，近处有毗邻的城市板球场和梅西大学（钟塔），远处是城市在向大海延伸，空阔无限。

4-031 Site observation and record (2), looking north from the south, Chinese Embassy in Wellington, New Zealand (20 October 2015)

Looking north from the southern end (the highest point) of the sloping land, one can see that No. 73 Rugby Street looks like a nail wedged into the site. The city's cricket pitch and Massey University (clock tower) are adjacent to the site. In the distance, the city is extending to the sea, wide and unlimited.

Who waches you?

'2015.10.20 Wellington 维多库亚地自由北望

4-032 延庆海坨山，延庆赛区二级输水泵站场地（之一）

（2017.12.08）

萧瑟寒山。

4-032 Site of the Secondary Water Pumping Station in Yanqing Zone, Haituo Mountain, Yanqing (1) (8 December 2017)

Bleak and cold mountains.

2017.12.08　延庆海坨二级泵站场地

4-033 延庆海坨山，延庆赛区二级输水泵站场地（之二）
（2017.12.08）

　　山林中废弃的农家乐餐厅，有鱼池在侧，有山溪在前，水冻结成冰。

4-033 Site of the Secondary Water Pumping Station in Yanqing Zone, Haituo Mountain, Yanqing (2) (8 December 2017)

　　The abandoned farmhouse restaurant in the mountain and forest has a fish pond on one side and a mountain stream in front. The water freezes into ice.

4-034 延庆海坨山，延庆赛区二级输水泵站设计构想

（2017.12.08）

一桥（车行）傍池，一桥（人行）跨溪。建筑化整为零，组构庭院。倚山临溪，坐落石造台基之上之侧，又有檐下屋顶平台，可居高望远。如山居聚落，又是市政公共设施。

在北京冬奥会延庆赛区，我们（不得不）做了很多的市政设施——缆车站、变电站、水泵站、造雪机房、气象雷达站、管廊出入口，等等。从建筑尺度、形象到结构和内外空间，城市市政设施的公共性设计探索，朝阳区高安屯垃圾处理／焚烧／发电中心是个起点，很有挑战，很有意义，也很有意思。

2017.12.08 延庆海坨二级输水泵站

4-034 Design concept of the Secondary Water Pumping Station in Yanqing Zone, Haituo Mountain, Yanqing (8 December 2017)

One bridge (for cars) is along a pool, and the other (for pedestrians) crosses the stream. The station is divided into several courtyard buildings. Against the mountain and near the stream, it is located on the side of a stone foundation and has a platform covered by a roof under which people can view the scenery. It is like mountain settlements, but municipal facilities.

In the Yanqing Zone of the Beijing Winter Olympics, we had to design a lot of municipal facilities: cable car stations, substations, pump stations, snow machine rooms, weather radar stations, pipe rack entrances and so on. In terms of the architectural scale, image, structure and internal and external space, the Incineration Center of Waste Comprehensive Treatment Plant at Chaoyang District, Beijing is the starting point for exploring the public nature of the urban municipal facilities. This is very challenging, meaningful and interesting.

4-035 延庆，小海坨山，北京冬奥会延庆赛区（高山滑雪中心）场地踏勘 GPS 轨迹图（2016.09.22）

特殊的草图——由飞往小海坨山顶的直升机航线和人的徒步行走路径组"绘"而成。海拔高程落差 1300 余米，中间有一段，是徒步重复爬上爬下的一次。

4-035 GPS track map of the site survey of the National Alpine Ski Centre, Beijing Winter Olympics Yanqing Zone, Xiaohaituo Mountain, Yanqing (22 September 2016)

A special sketch—the GPS track "drawing" consists of a helicopter route to the top of the Xiaohaituo Mountain and people's walking path. The altitude difference is more than 1,300 metres. In the middle section, one must climb up and down on foot.

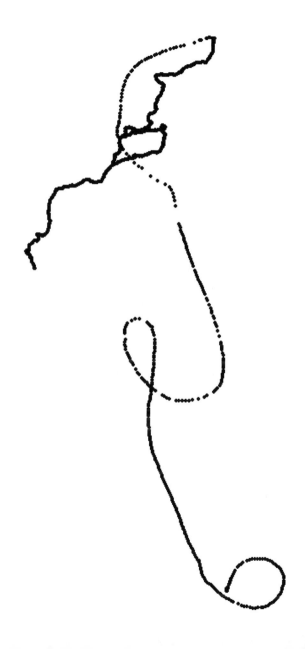

4-036 延庆，小海坨，山顶俯瞰整个赛区场地（2016.09.21）

只有在山中行走，才能真正理解山。如果不用手脚、眼睛和内心亲身体验真正的自然山水，怎么能产生与山水关联的想象和营造呢？

4-036 Overlooking the entire site of the zone from the top of the mountain, Xiaohaituo Mountain, Yanqing (21 September 2016)

Only by walking in the mountains can one truly understand them. If you do not experience the true natural landscape with your hands, feet, eyes and heart, how can you create the images and objects associated with the landscape?

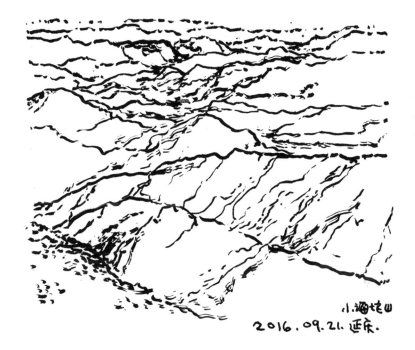

小淘坑山
2016.09.21. 延庆.

4-037 延庆，由气象站观察点南望小海坨山谷（2016.05.10）

4-037 Looking south to the Xiaohaituo Valley from the observation point of the weather station, Yanqing (10 May 2016)

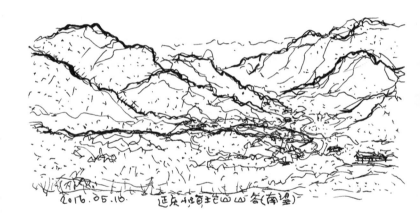

2016.05.10. 延庆小海陀主峰山山谷(南望)

4-038 延庆海坨山，北京冬奥会延庆赛区冬奥村及赛区入口安检广场设计构想（2017.03.16）

由气象站望延庆冬奥村及小海坨方向。山林掩映中的台地建筑和景观。

4-038 Design concept of the Olympic Village and the entrance square of the security checkpoint in the Yanqing Zone of the Beijing Winter Olympics, Haituo Mountain, Yanqing (16 March 2017)

Looking at the Yanqing Olympic Village and Xiaohaituo Mountain from the Weather Station. Terraced buildings and landscape are found in the mountains and forests.

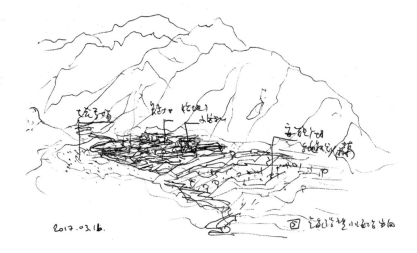

2017.03.16.

4-039 延庆海坨山，延庆冬奥村场地中的小庄科村遗迹

（2018.07.03）

山林掩映。

4-039 Ruins of Xiaozhuangke Village on the site of the Yanqing
Olympic Village, Haituo Mountain, Yanqing (3 July 2018)

The ruins are shaded by mountain and forest.

2018.07.03. 延庆冬奥村小庄科村 遗址

4-040 延庆海坨山，由休息点 A 望雪车雪橇中心"男人址"山坡（S2）（2016.08.30）

选址何处？

"男人坡"硬朗而凸显于群山中。

4-040 Looking at the "man-site" hillside (S2) of the National Sliding Centre from rest point A, Haituo Mountain, Yanqing (30 August 2016)

Where is the site?

The "man-site" hillside is tough and prominent in the mountains.

2016.08.30 匡氏   VIEW FROM THE COFFEE POINT (A)

4-041 延庆海坨山，由休息点 B 望雪车雪橇中心 "男人址" 山坡（S2）（2016.08.30）

4-041 Looking at the "man-site" hillside (S2) of the National Sliding Centre from rest point B (S2), Haituo Mountain, Yanqing (30 August 2016)

2016.08.30 延庆

VIEW FROM THE COFFEE POINT (CB)

4-042 延庆海坨山，由气象站看雪车雪橇中心"男人址"山坡（S2）（2016.08.30）

4-042 Looking at the "man-site" hillside (S2) of the National Sliding Centre from the weather station, Haituo Mountain, Yanqing (30 August 2016)

2016.08.30 · 延庆　　　VIEW FROM WEATHER OBSERVATION STATION

4-043 延庆，由雪车雪橇中心"男人址"山坡（S2）出发点（1020 高程）望气象站及群山（2016.08.30）

4-043 Looking at the weather station and the mountains from the starting point of the "man-site" hillside (S2) of the National Sliding Centre (the elevation is 1020 metres), Yanqing (30 August 2016)

671

2016.08.30. 延庆

VIEW FROM 1020 START POINT

4-044 延庆海坨山，由观察点望雪车雪橇中心"女人址"山谷（E1）（2016.08.31）

与"男人址"不同，"女人址"如此柔软，隐藏于山谷之中，被树木覆盖和遮阴，一条溪流在中间向下流淌。

4-044 Looking at the "woman-site" valley (E1) of the National Sliding Centre from the observation point, Haituo Mountain, Yanqing (31 August 2016)

Unlike the "man-site", the "woman-site" is softly hidden in the valley, covered and shaded by trees, and a stream flows down in the middle.

2016.08.31 延庆

VIEW FROM VIEW POINT TO E1

4-045 延庆海坨山，北京冬奥会延庆赛区国家雪车雪橇中心，现场构想草图（之一）（2017.03.16）

　　由小平台南望雪车雪橇中心。

4-045 On-site conceptual sketch of the National Sliding Centre in the Yanqing Zone of the Beijing Winter Olympics, Haituo Mountain, Yanqing (1) (16 March 2017)

　　Looking south towards the National Sliding Centre from the small platform.

2017.03.16.

百 小龙台瞭望

4-046 延庆海坨山，国家雪车雪橇中心，现场构想草图（之二）（2017.03.16）

由观景台（气象站）望山坡上的雪车雪橇中心。

4-046 On-site conceptual sketch of the National Sliding Centre in the Yanqing Zone of the Beijing Winter Olympics, Haituo Mountain, Yanqing (2) (16 March 2017)

Looking at the National Sliding Centre from the observation deck (weather station).

2017.03.16 　　　　　　　欢喜岭望祁连 ⑰

4-047 延庆海坨山，北京冬奥会延庆赛区，现场构想草图（之一）（2017.06.16）

由休息点眺望冬奥村（左）和雪车雪橇中心（右）。

4-047 On-site conceptual sketch of Yanqing Zone of the Beijing Winter Olympics, Haituo Mountain, Yanqing (1) (16 June 2017)

Viewing the Olympic Village (left) and the National Sliding Centre (right) from the rest point.

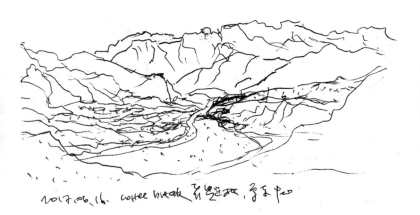

4-048 延庆海坨山，北京冬奥会延庆赛区，现场构想草图（之二）（2017.06.16）

由小海坨中间平台向西南方向眺望山下（冬奥村和雪车雪橇中心）。

4-048 On-site conceptual sketch of Yanqing Zone of the Beijing Winter Olympics, Haituo Mountain, Yanqing (2) (16 June 2017)
Looking southwest and downhill (the Olympic Village and National Sliding Centre) from the central platform of the Xiaohaituo Mountain.

4-049 延庆，小海坨，北京冬奥会延庆赛区国家高山滑雪中心，现场草图（之一）（2017.06.16）

　　多次转弯的赛道，贴在陡坡的滑雪"廊道"，山脊上的树林和山坡上的大树是不同位置赛道的特征标志物，人工的赛道与自然的树木和山坡互动相成。

4-049 On-site conceptual sketch of the National Winter Alpine Ski Centre, Yanqing Zone of the Beijing Winter Olympics, Xiaohaituo Mountain, Yanqing (1) (16 June 2017)

　　The track with many turns and the ski "corridor" are stuck on the steep slope. The woods on the ridge and the big trees on the hillside are markers on the track at different positions. The man-made track interacts with the natural trees and hillside.

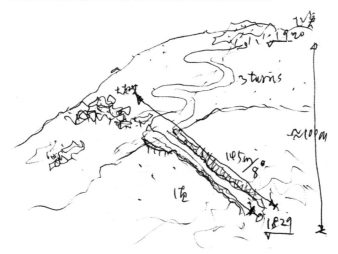

4-050 延庆，小海坨，国家高山滑雪中心，现场草图（之二）
（2017.06.16）

山顶平台，滑降赛道，回转和大回转赛道。

4-050 On-site conceptual sketch of the National Winter Alpine Ski Centre, Yanqing Zone of the Beijing Winter Olympics, Xiaohaituo Mountain, Yanqing (2) (16 June 2017)

Platform on the peak, down-hill ski course, slalom and giant slalom ski courses.

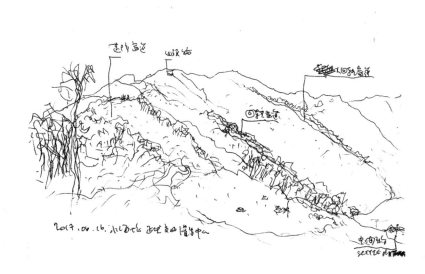

4-051 延庆，小海坨，国家高山滑雪中心，现场草图（之三）（2017.06.16）

结束区设计构想。

4-051 On-site conceptual sketch of the National Winter Alpine Ski Centre, Yanqing Zone of the Beijing Winter Olympics, Xiaohaituo Mountain, Yanqing (3) (16 June 2017)

Design concept of the end zone.

683

2017.06.16.

1382

4-052, 053, 054, 055

延庆，小海坨，国家高山滑雪中心，现场草图（之四、五、六、七）（2017.06.16）

　　想象飞速滑行中的动景。滑行者由中间平台下行山谷，再转弯过来，视野中的"鲁西石"好像出现在赛道的中间，"像廊道一样保持赛道的狭窄感"。（伯纳德·鲁西语）

2017.06.16.

4-052, 053, 054, 055 On-site conceptual sketch of the National Winter Alpine Ski Centre, Yanqing Zone of the Beijing Winter Olympics, Xiaohaituo Mountain, Yanqing (4, 5, 6, 7) (16 June 2017)

Imagining the "moving scene" of fast skiing. From the middle platform, down to the valley and turning around again, the skier can see that Russi Rock seems to appear in the middle of the ski course. Keep it narrow as a gallery, says Bernard Russi.

2017.
06.16   B. Russi Rock in the middle
        = keep it narrow as a gallery"
        (on se the terrain)

2017.06.16

4-056, 057, 058

延庆，小海坨，国家高山滑雪中心，山顶出发区草图（之一、二、三）（2016.09.21—22, 2019.01.14）

　　海拔 2198 米的小海坨山顶下方的"大风筝"，后部和两侧掩埋在山体之中。

4-056, 057, 058 Sketches of the departure area on the peak, National Alpine Ski Centre, Xiaohaituo Mountain, Yanqing (1, 2, 3) (21-22 September 2016, 14 January 2019)

　　Below the peak of the Xiaohaituo Mountain (at an altitude of 2198 meters), the back and both sides of the "large kite" are buried in the mountain.

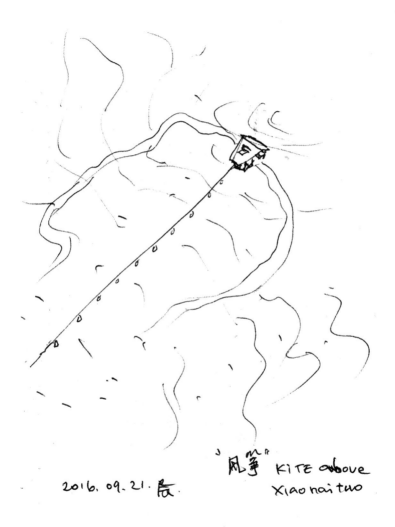

2016. 09. 21. 辰.

"风筝" KITE above
Xiao nai tuo

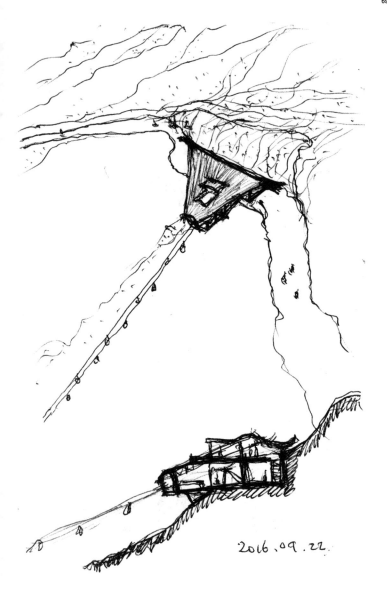

2016.09.22.

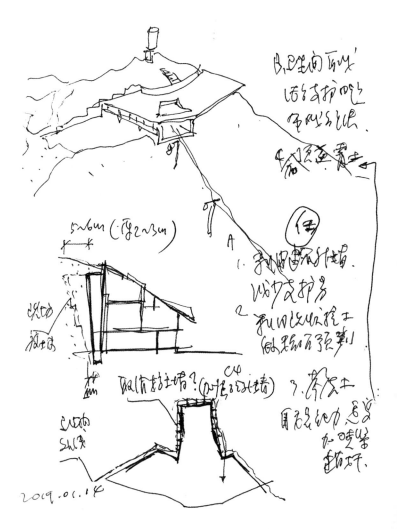

4-059 延庆，北京冬奥会延庆赛区，全区草图（2017.07.22）
立轴山水。山林场馆，生态冬奥。

4-059 Sketch of the whole zone, Yanqing Zone of the Beijing Winter Olympics, Yanqing (22 July 2017)

Vertical axis landscape painting. Mountain Forest Venues, Ecological Winter Olympic Games.

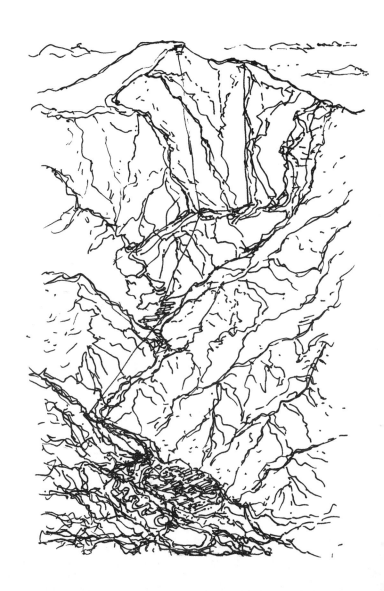

4-060 正蓝旗，元上都遗址博物馆，现场构思草图
（2009.03.14）

4-060 On-site sketches of the design concept, the Museum for Site of Xanadu, Zhenglan Banner (of Inner Mongolia) (14 March 2009).

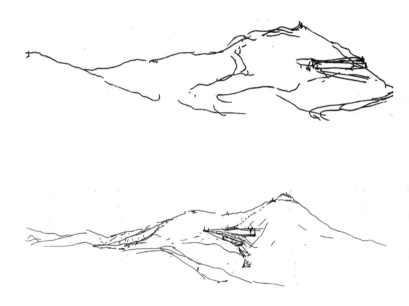

4-061 崇礼，太子城，崇台——北京冬奥会张家口赛区冬奥展示中心，设计草图（之一）（2019.05.22）

崇台与太子城遗址中轴线和太子城雪花小镇。

沿崇台纵轴方向延伸，在与太子城遗址中轴线交汇处，恰是雪花小镇文创商街的北端起点和入口。

4-061 Design sketch of Chongtai—Winter Olympic Exhibition Centre, Zhangjiakou Zone, Taizi Town, Chongli (1) (22 May 2019)

Chongtai, the central axis of the Taizi Town ruins and the snowflake town of Taizi Town.

The point where the extension of Chongtai's vertical axis intersects the central axis of Taizi Town's ruin site is the starting point and entrance to the northern end of the cultural and commercial street in the snowflake town.

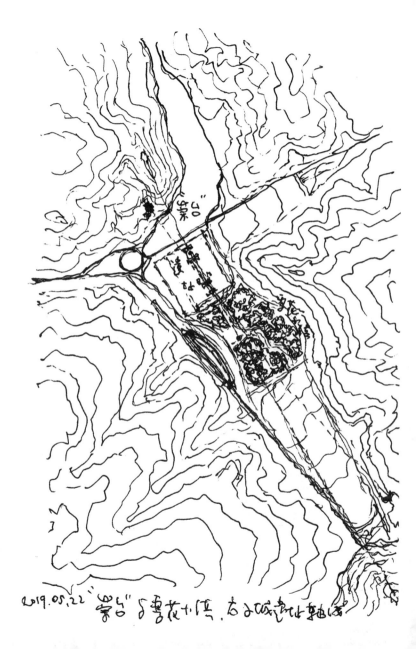

2019.05.22 "瞿昙" δ青花小洋，古み玻璃窗北轴汤

4-062 崇礼，太子城，崇台——北京冬奥会张家口赛区冬奥展示中心，设计草图（之二）（2019.05.22）

以山为宗。

4-062 Design sketch of Chongtai—Winter Olympic Exhibition
Centre, Zhangjiakou Zone, Taizi Town, Chongli (2) (22 May 2019)
Mountain as a key reference for design.

4-063 崇礼，太子城，崇台——北京冬奥会张家口赛区冬奥展示中心，设计草图（之三）（2019.05.22）

崇台，崇礼之台，宗山之台也。

4-063 Design sketch of Chongtai—Winter Olympic Exhibition Centre, Zhangjiakou Zone, Taizi Town, Chongli (3) (22 May 2019)

Chongtai, the platform of Chongli and the platform embodying an attitude to revering mountains.

2019.05.22. 紫荆太和城"景台"。紫荆之名也，案山之台也。

4-064 崇礼，太子城，崇台——北京冬奥会张家口赛区冬奥展示中心，设计草图（之四）（2019.05.22）

左：近山者，折也；右：远山者，曲也。

由西南侧下方山路仰视，隐在半山又悬挑浮出的崇台。

4-064 Design sketch of Chongtai—Winter Olympic Exhibition Centre, Zhangjiakou Zone, Taizi Town, Chongli (4) (22 May 2019)

The left side of the building, near the mountain, is folded, whereas the right side, away from the mountain, is curved.

Looking up to the cantilevered Chongtai, hidden in the middle of the mountain, from the mountain road below the southwest side.

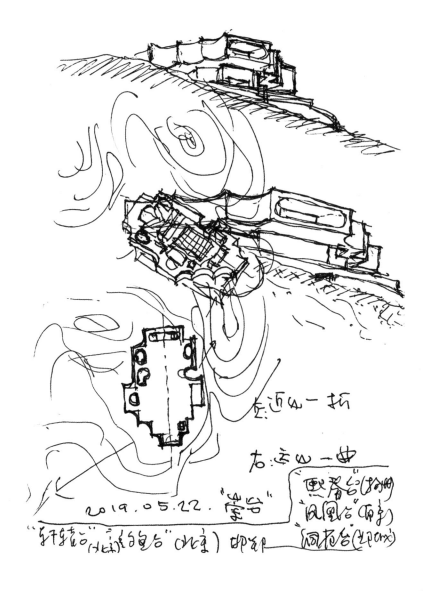

左近山一拓

右远山一曲

默卷台(拟拟)
凤凰台(雨别)
洞府台(北咖x)

2019.05.22. "箅台"

"轩辕台"(北蓝)"钓鱼台"(北蓝) 邯郸

4-065 崇礼，太子城，崇台——北京冬奥会张家口赛区冬奥展示中心，设计草图（之五）（2019.05.27）

由引道入口外看崇台全貌，自由伸展的台基嵌在山坡，水平体量悬浮在台基的上空。

4-065 Design sketch of Chongtai—Winter Olympic Exhibition Centre, Zhangjiakou Zone, Taizi Town, Chongli (5) (22 May 2019)

Looking at the totality of Chongtai from the entrance to the approach road. The freely extending platform is embedded in the hillside, and the horizontal volume of Chongtai is suspended in the air above the platform.

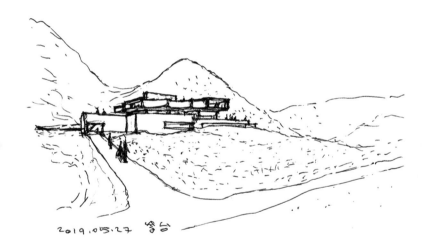

2019.05.27 姿白

4-066 崇礼，太子城，崇台——北京冬奥会张家口赛区冬奥展示中心，设计草图（之六）（2019.05.27）

崇台西入口前区透视，台基与"浮台"。

4-066 Design sketch of Chongtai—Winter Olympic Exhibition Centre, Zhangjiakou Zone, Taizi Town, Chongli (6) (22 May 2019)

Perspective of the front area of the west entrance of Chongtai, base and floating platform.

2019.05.27 烟台

4-067 廊坊，华夏幸福幼儿园建筑与景观意象草图
（2016.03.20）

造园。山与房。

4-067 Image sketch of Huaxia Happiness Kindergarten's
architecture and landscape, Langfang (20 March 2016)
Gardening. Mountain and house.

4-068 浙江台州，玉环博物馆和图书馆，设计草图
（2015.05.29）

在填海造地的新城，再现坎门渔港空间胜景。

结构／空间单元在水平、垂直两个向度的组合。聚落的营造。

4-068 Design sketch of Yuhuan Museum and Library, Taizhou, Zhejiang Province (29 May 2015)

In the new city where the land is reclaimed, the Kanmen fishing port space of poetic scenery will be reproduced.

The combination of structure/space unit in two dimensions, horizontally and vertically. The design and construction of a settlement.

711

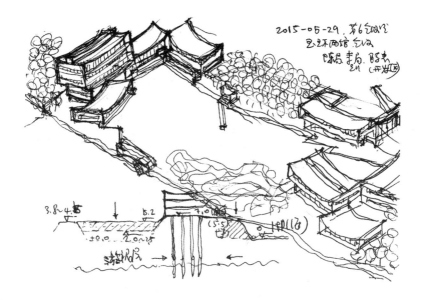

2015-05-29. 第6会议室
乌江画馆 512
陈丹青 题字 (开发区)

3.8~4.5
6.2
±0.10
(5.5)
±0.0   2.0~2.5

4-069 河南，商丘博物馆，设计草图（2009.02）

关于"城市"与"建筑"的同构、复合及其概念转化，我有着特别而长久的兴趣，曾经做过多次尝试（从大学本科毕业设计开始），商丘博物馆是我第一次把这个意图落地实施的项目——以现存归德古城为代表的黄泛古城形制为蓝本，转化为新建筑的内外空间构成，并借堤台景观营造回游路径，空间内外观游一体，使得这个房子成为城—筑—园的混合体，这种"混合"也很能代表我们这些年研究、思考、实践的特点。这个项目历时6年，数易其稿，酒也为它喝了不少，仍有诸多不如意之处（特别是展陈设计的失控），但在我心里有特殊的位置。

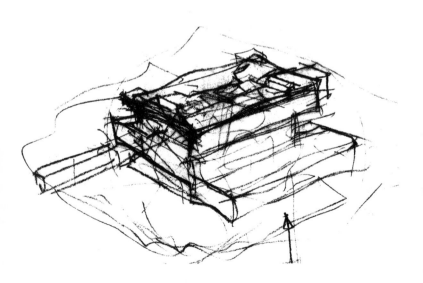

4-069 Design sketch of Shangqiu Museum, Henan Province (February 2009)

I have a special and long-lasting interest in the isomorphism and integration of the city and buildings and their conceptual transformations. I have tried many times (starting from my undergraduate final year of design), and the Shangqiu Museum is the first built project implemented with this intention. The inner and outer space of the new building has been transformed from the existing ancient city in the Huangfan area (affected by the Yellow River flooding) represented by the city of Guide. The architects created a wandering path by using the embankment landscape. Both inside and outside space, exhibition and wandering routes are integrated, making the museum a mixture of the city, building and garden. This mixture also represents the characteristics of our research, thinking and practice over the years. This project lasted six years, and numerous proposals were produced in the process. I also drank a lot of wine with clients to get this done. Despite the many unsatisfactory things about it (especially the lack of control over the exhibition design), it holds a special place in my heart.

4-070 唐山，第三空间综合体，设计草图（2016.09.28）

陪金秋野老师又访唐山第三空间，言及柯布及其"别墅大厦"方案，忽然想到此处每层人工台地之上的日常居游空间，意图更靠近"辋川别业"之别业宅园，即此"别墅"非彼"别墅"（Villa）也，所以第三空间可称是层层叠摞的"别业大厦"了。又及登至屋顶，脚下水平排列的城市住宅，与眼前竖向分布、密匝错落的亭台小屋形成鲜明对照，大坡屋顶上的天窗也似一群山坡小屋，仿佛聚落随着土地减缩，而由平原向高地的自然延伸和生长。

4-070 Design sketch of the "Third Space" in Tangshan (28 September 2016)

Accompanying Jin Qiuye, I visited the Third Space Complex in Tangshan again, talking about Le Corbusier and his Plan Voisin, and suddenly thinking of the daily living space of stepping floors here. My intention was closer to the home-garden of the Wangchuan Villa (*Wangchuan bieye*). The "Third Space" can be called a Villa Mansion. Looking out from the roof, the urban houses arranged horizontally under the feet are in sharp contrast to the building's pavilions, that are vertically, densely and disorderly distributed on the facade. The skylights on the big pitched roof are also like a group of hillside huts; as if with the shrinking of the land the human settlements naturally extended and grew from the plain to the highland.

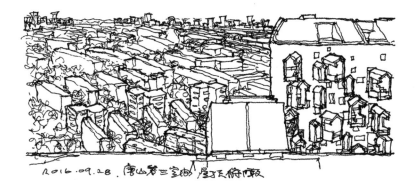

2016.09.28. 唐山第三宾馆 屋顶天俯瞰

4-071 北京，仓阁——首钢工舍，设计草图 （2016.01.04）

废墟"自然"：2015 年格物工作营南京花露岗"瞬时桃花源"中的"台阁"在北京石景山的建筑放大版。

4-071 Design sketch of "Silo Pavilion", Holiday Inn Express Beijing Shougang (4 January 2016)

The ruined "nature": the Terrace Pavilion, appearing in the Instantaneous Peach Garden project in 2015 Nanjing Hualugang workshop of *Gewu* was enlarged and became a building in Shijingshan District, Beijing.

4-072 广州，某足球场设计草图（2019.06.21—24）

"浮云"：草坡基座＋浮云球场＋悬浮屋顶。

4-072 Design sketch of a football stadium, Guangzhou（21-24 June 2019）

"Floating clouds".

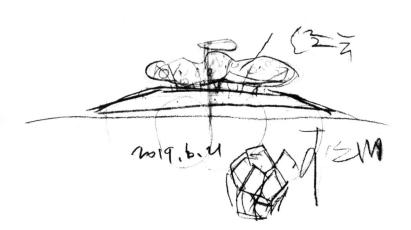

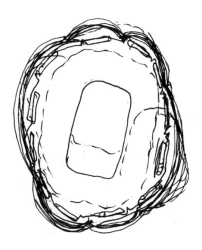

2019.06.24.

4-073 北京，大院胡同 28 号院改造，设计草图（2017.02）

微缩北京。旧城新作，大都小作，小题大作。

虽然身处"大院"，其实是个小院，虽是小院，却有些大意图：其一，想把这个小院改造做成解决北京旧城更新"三道难题"的实验版；其二，从微缩宅园、微缩社区到"微缩北京"，从宅园合一到都市胜景，这个小院项目，虽然尚显粗简，但把我从 1990 年夏天爬上景山看故宫开始的对中国城市、建筑、园林、聚落的兴趣、研究和思考都融汇在一起，这甚至是我自己都始料未及的。也难怪这么小的项目，居然写出那么多字的文章来，思考的核心是在呼应一种影响至深至久至远的文化精神，它由不同类型、不同层次的中国营造传统，反复地、系统地透露和强调，渗透到人们的内心和生活深处。

大院小院，秋色秋意。期待四季轮转，生活长在。

4-073 Design sketch of Renovation of No. 28 Dayuan Hutong, Beijing (February 2017)

Miniature Beijing. It is a new work in the old city, a small project in the big city, and a big endeavour to address small issues. Although it is located in a "big courtyard" (*dayuan*), the building is actually a small courtyard with some big intentions. First, I want to transform this small courtyard house into an experimental version that solves the three difficulties of Beijing's old city renewal. Second, from miniature home-gardens, miniature communities to a miniature Beijing, from home-garden integration to an urban poetic scenery, this small courtyard project, although still rough, combines my interest, research and thinking about Chinese cities, buildings, gardens and settlements, since I climbed up Jingshan Hill to see the Forbidden City in the summer of 1990. This is even unexpected for me. It is no wonder that for such a small project I actually wrote so many words. The core of my thinking was to echo a cultural spirit that has a deep, far and long-lasting influence. It is created by different types and levels of Chinese building tradition, repeatedly and systematically revealed and emphasised, penetrating people's hearts and lives.

The small courtyard in the *dayuan* is full of colour and images of autumn. Looking forward to the rotation of the four seasons and to the continuity of life.

大陆胡同28号院已改造"徽州府朝堂"

4-074 北京，壁园，平面及轴测草图（2018.04.18）
壁山，壁廊，壁亭，见山轩。

4-074 Plan and axonometric sketch of Biyuan Garden, Beijing (18 April 2018)

Wall mountain, wall gallery, wall pavilion, Jianshanxuan Pavilion.

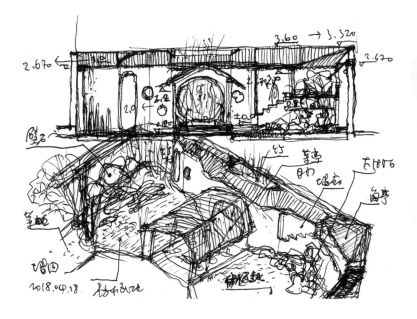

2018.04.18

4-075 北京，壁园，平面及构造草图（2018.05.11）

轩移一侧，中加木桥，位置经营。

4-075 Plan and detail sketches of Biyuan Garden, Beijing (11 May 2018)

The pavilion was moved to one side, and a wooden bridge was added to the centre. The design process is a process of operating position (*weizhi jingying*).

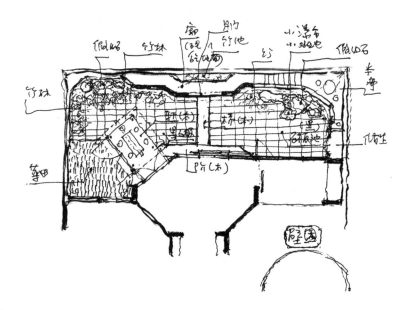

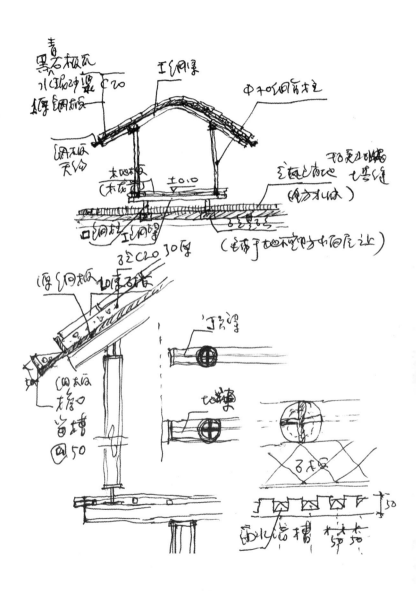

4-076 北京，璧园，透视草图 （2019.06.04）

4-076 Perspective sketch of Biyuan Garden, Beijing (4 June 2019)

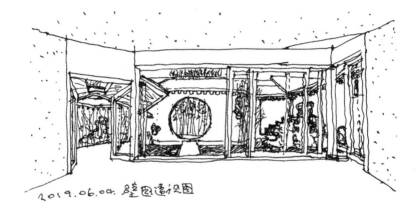

2019.06.04. 壁园透视图

4-077 北京，璧园，假山草图两幅 (2019.06.04)

掇山叠石：是谁"鬼斧神工"？湖石局部切削与组构——人工与自然的"造化"。

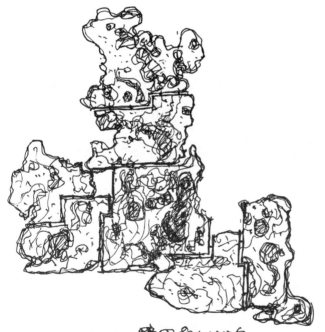

2019.06.04 璧园登山机会
太湖石切削组合
"人工+风吃"="叠山"

4-077 Two sketches of the rockery in Biyuan Garden, Beijing (4 June 2019)

Making rockery and stacking stone (*duoshandieshi*): who is extremely skillful? Partially cutting and composing of the lake stone is a creation made by man and nature.

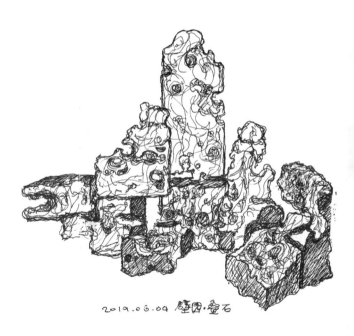

2019.06.04 壁园·叠石

4-078 上海，聚落套椅，草图 （2014.09）

　　"聚落"（set-all）：聚而落座。一组高低错落具有空间感的椅子群。外框内实，大小、深浅不一，人与椅一起形成别样的坐姿图景。鲁安东兄评说："坐的地形"，有触动。或许是潜意识的，做的时候还真没这么想，只是按酒吧的"地形"和人的行为动作做了大小、高低的对应。这么说来，还有"坐的朝向""坐的空间""坐的场所"等等，看来家具和"坐"还真的是没那么简单。卢永毅老师说，在中国的传统中，家具是城市和建筑手法的延伸，像一座座"微小的房子"，容纳人的身体与生活。

4-078 Sketch of set-all chairs, Shanghai (September 2014)

　　The set-all: a group of chairs with different heights and spatial senses are used for people to gather and sit. The frame is in the outside and the inside is solid. The sizes and depths are different. People and chairs together form an alternative scene of sitting posture. Lu Andong's commentary of "terrain of sitting" is inspirational. Perhaps it is a subconscious, as when I did it. I just made the size and height of the chairs according to the "terrain" of the bar and people's behavior. In this way, there are also sitting orientation, sitting space, sitting place, etc. It seems that furniture and "sitting" are really not that simple. Lu Yongyi said that in the Chinese tradition, furniture, as an extension of the city and buildings, is like a "small house" that accommodates the body and life.

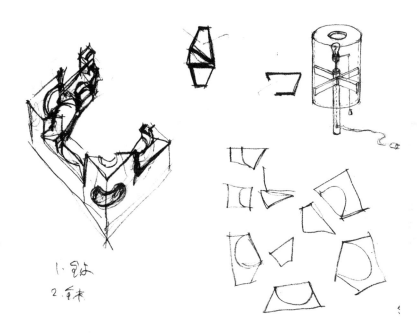

1. 身体
2. 细木

4-079 胜景城市 （2019.11.22）

4-079 The city of poetic scenery (22 November 2019)

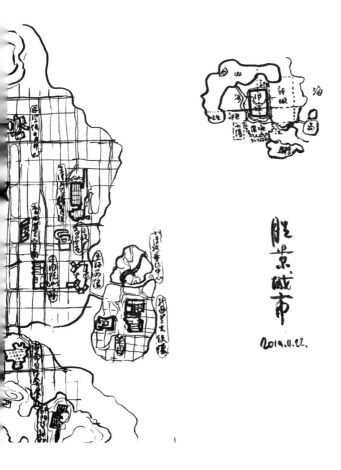

4-080 绘五年前高铁上所见：北方，冬季，雪后，田野，村庄（2019.01.01）

与五年前此时同感：2019，一个新的开始。期待生活工作有进步。思考，劳作。从容，不迫。清明，喜悦。

4-080 Drawing of what I have seen on the high-speed train five years ago: north, winter, snow, fields, villages (1 January 2019)

With the same feeling that I had five years ago, the year 2019 is a new beginning. Looking forward to progress in life and work. Thinking, working, keeping calm and patient. Clear and bright, joy.

2019. 元旦 录2012年底图：很爱这北方的冬天. 萧瑟. 静默, 内在着力量。

后记

　　观、想、做三位一体。这本小册子与同期出版的另外两本书呈现出关联"互引"的关系：《胜景几何论稿》是思考之"述"，《李兴钢 2001—2020》则是实践之"作"，本书作为行观之"悟"，甚而也可以被视作"胜景几何"思想形成和实践操作背后的思考图示与密码。在大学时代，绘图或者绘制草图是建筑学生必须掌握的"基本功"之一。作为一种建筑设计的绘图表现手段，彭一刚先生的经典"小钢笔"图与周恺老师的白描墨线图，具有传承和关联的两种代表风格，都对我产生了不小的影响，我也是他们喜爱并擅长"画图"的学生之一。但大学毕业后从事建筑师工作的很长时间里，绘制草图主要是作为设计思考的工具，我也始终未将"绘图"向"绘画"的方向发展，似乎已经基本放下了这项"技艺"。2014 年夏天在晋东南古建筑考察中，我将彼时彼刻类似当年爬景山看故宫时的强烈触动总结为"晋东南五点"，并分别绘制一幅对应的草图，发在刚开通不久的微信朋友圈。从此一发不可收，由之后的蔚县村堡、五台山佛光寺到北京颐和园、无锡寄畅园，由山水仿画到真山大水，由龙门石窟、徽州古村到自构太行小五台"不断展开的山水"，再到延庆海坨山中的"高山流水""山林场馆"，游、行、望、居，憩、旅、栖、造，直至收获这册《行者图语》。草图不再仅是设计的草图，还是我在自然的现场所记录的即时心得，并伴以文字，如同古人的册页般，成为由"图"和"语"共同构成的旅途和心路日记。感谢张永和老师无比忙碌中欣然应允为这本小册作序，并提供宝贵的提示和建议；感谢所有"旅途"中的组织方、导引者和同行人、陪伴者，他们也可以说是这些"图语"的参与和见证者；感谢工作室出版

Acknowledgements

Envisioning, thinking, and making are the Trinity. This booklet and the other two books published over the same period show the relationship of "cross-references": *Essays about Integrated Geometry and Poetic Scenery* is the "description" of ideas, and *Li Xingggang 2001—2020* is the "work" of practice. As the "realization" of the observations during travel, I regard this book as the thinking diagram and thought code behind the formation and practical operation of the "Integrated Geometry and Poetic Scenery" ideas. In my university period, mastery of drawing or sketching, as a representation method of architectural design, was one of the "basic skills" of architecture students. Academician Peng Yigang's classic" small pen" drawings and Professor Zhou Kai's white ink lines represent two styles with inner inheritance and connection that have constantly influenced me. I am, undoubtedly, one of the students good at "drawing" that they love. For a long time working as an architect after graduation, however, I used drawing, especially sketching, mainly as a means of design thinking and expression. I was never interested in developing "drawing" towards "painting." It seemed that I had basically neglected this "craft". However, in the summer of 2014, during a survey of ancient buildings in the southeast Shanxi Province, I summarised the strong impressions I felt similar to the feelings when climbing the Jingshan Hill to see the Forbidden City as "the five points of southeastern Shanxi Province", drew five corresponding sketches, and posted them on my WeChat Moments (a function recently launched at that time). Since then, I have drawn and chosen these sketches and these words for this book. Structured into four sections—wandering, walking, viewing, and living, it documents a wide range of experiences and works, from a village fortress in Yu County, the Foguang Temple

740

小组伙伴们的整理、编辑和排版，这是一项烦琐无比的工作；感谢丁光辉老师的英文翻译和浙江出版联合集团的出版发行，使我的这些个人化"涂写"可以与读者们分享。感谢我的父母、妻儿，他们永远是我漫漫建筑旅程之后回归的港湾。

李兴钢

二〇二〇年二月于北京

in Wutai Mountain, the Summer Palace in Beijing, to the Jichang Garden in Wuxi; from landscape paintings to the real Huangshan Mountain and Yangtze River; from the Longmen Grottoes, ancient villages in Anhui Province to the self-composed "continuously unfolding landscape" of Taihang Xiaowutai Mountain; from "mountains and forest venues" in the Haituo Mountain in Yanqing, to the scenes of the 2022 Olympic Winter Games. Sketches are no longer just preparations for design, but have become records of my real-time experiences of the scene where humans commune with nature. My sketches and the words—similar to ancient albums of paintings or calligraphy—are diaries of my physical and intellectual journeys. I would like to thank Yung Ho Chang for his willingness to write a forward for this booklet and for his generosity in providing valuable tips and suggestions. Thanks go to all the organisers, guides, peers, and companions during the "trips." They are the participants and witnesses of these drawings and phases. I appreciate my studio's publication group members for taking on the tedious work of collation, editing, and layout. I am also grateful for Ding Guanghui's English translation and for the publication and distribution efforts of Zhejiang Publishing Group, which enable my personal "drawings and writings" to be shared with a wide range of readers at home and abroad. Thanks to my parents, wife, and son, who are always home waiting for my return after long architectural journeys.

Li Xinggang
Beijing, February 2020

　　李兴钢，1969 年出生于中国唐山，1991 年和 2012 年分别获得天津大学学士和工学博士学位，2003 年创立中国建筑设计研究院李兴钢建筑工作室，现任中国建筑设计研究院总建筑师、天津大学客座教授 / 博士生导师和清华大学建筑学院设计导师。 以"胜景几何"理念为核心，建筑研究与实践聚焦建筑对于自然和人密切交互关系的营造，体现独特的文化厚度和美学感染力。获得的国内外重要建筑奖项包括：亚洲建筑师协会建筑金奖(2019)、ArchDaily 全球年度建筑大奖（2018）、WA 中国建筑奖（2014/2016/2018）、全国优秀工程设计金 / 银奖(2009/2000/2010)等,是中国青年科技奖(2007)和全国工程勘察设计大师荣誉称号（2016）的获得者。作品参加威尼斯国际建筑双年展（2008）等国内外重要展览，并于北京举办"胜景几何"作品个展（2013）。

Li Xinggang received his Doctor of Engineering degree from Tianjin University and founded the Atelier Li Xinggang in 2003. He is visiting professor of Tianjin University and design tutor at the School of Architecture, Tsinghua University. His architectural practice and research focus on the idea of "Integrated Geometry and Poetic Scenery", emphasizing the cultural depth and aesthetic affection in the close reaction of architecture to nature and human-beings. His practice was honored with design awards including the Gold Award of ARCASIA Awards for Architecture (2019), the WA Chinese Architecture Awards (2014, 2016, and 2018) and ArchDaily Building of the Year Award (2018), Gold and Silver awards of the National Excellent Engineering Investigation and Design Awards, etc. He also received the China Youth Science and Technology Award (2007) and the esteemed National Engineering Survey and Design Master Award (2016). He took part in major architecture exhibitions in China and abroad, including Venice Architecture Biennale (2008), etc., and held his architecture solo exhibition "Integrated Geometry and Poetic Scenery" in Beijing (2013).

出 版 人：徐凤安
责任编辑：王　巍　金慕颜
责任校对：高余朵
责任印刷：汪立峰

装帧设计：姜汶林
李兴钢建筑工作室团队：
　　姜汶林、侯新觉、孔祥惠
英文翻译：丁光辉
英文审校：吴介邦（Kenneth Wu）

图书在版编目（CIP）数据

行者图语 / 李兴钢著 .-- 杭州：浙江摄影出版社，
2020.9
　ISBN 978-7-5514-2765-4

　Ⅰ .①行… Ⅱ .①李… Ⅲ .①建筑设计－作品集－中
国－现代 Ⅳ .① TU206

中国版本图书馆 CIP 数据核字 (2019) 第 277311 号

XINGZHE TUYU

行者图语
李兴钢 著

全国百佳图书出版单位
浙江摄影出版社出版发行
　　　地址：杭州市体育场路 347 号
　　　邮政编码：310006
　　　电话：0571-85151082
　　　网址：www.photo.zjcb.com
经销：全国各地新华书店
印刷：北京卡梅尔彩印厂
开本：787mm×1092mm 1/32
印张：23.375
版次：2020 年 9 月第 1 版　2020 年 9 月第 1 次印刷
ISBN 978-7-5514-2765-4
定价：98.00 元
本书若有印装质量问题，请与本社发行部联系调换。